Conference Board of the Mathematical Sciences
REGIONAL CONFERENCE SERIES IN MATHEMATICS

supported by the
National Science Foundation

Number 72

INTRODUCTION TO ARRANGEMENTS

Peter Orlik

Published for the
Conference Board of the Mathematical Sciences
by the
American Mathematical Society
Providence, Rhode Island

Expository Lectures
from the CBMS Regional Conference
held at Northern Arizona University,
Flagstaff, Arizona
June 6–20, 1988

Research supported by National Science Foundation Grant DMS-8600408.

1980 *Mathematics Subject Classifications* (1985 *Revision*). Primary 05B35, 32C40, 57N65; Secondary 14F40, 14J99, 51A05.

Library of Congress Cataloging-in-Publication Data

Orlik, Peter, 1938-

Introduction to arrangements/Peter Orlik.

p. cm. —(Regional conference series in mathematics/Conference Board of the Mathematical Sciences; no. 72)

"Expository lectures from the CBMS regional conference held at Northern Arizona University, Flagstaff, Arizona, June 6–20, 1988"–T.p. verso.

Bibliography: p.

ISBN 0-8218-0723-4 (alk. paper)

1. Combinatorial geometry–Congresses. 2. Combinatorial enumeration problems–Congresses. 3. Lattice theory–Congresses. I. Conference Board of the Mathematical Sciences. II. Title. III. Series.

QA1.R33 no. 72

[QA167]

510 s—dc516/.13—dc19 89-14893

CIP

---

to my parents

Contents

Kirt

Daniel

Deodhar

List of Figures

Preface

An arrangement of hyperplanes is a finite collection of codimension one subspaces in a finite dimensional vector space over some field. Arrangements occur in several branches of mathematics: in the study of braids and phase transition, in wave fronts, in hypergeometric functions, in reflection groups and Lie algebras, in coding theory, in the study of certain singularities, in combinatorics and group theory, and in spline functions.

Some aspects of the theory have a distinguished history. This was reviewed in Grünbaum's book [60] and in his CBMS lectures [62]. Recent interest in the topological properties of the complement of an arrangement over the complex numbers started with papers by Arnold [3], Brieskorn [20], Deligne [33] and Hattori [64]. They studied the cohomology groups and the homotopy type of the complement. Orlik and Solomon [100] added combinatorial tools and Terao [139] used methods of algebraic geometry. These results were described by Cartier [24] in a Bourbaki seminar talk. These lecture notes provide an introduction to the new developments and survey the current activity in the area, with particular emphasis on the topological aspects. A more comprehensive treatment is forthcoming in a book written jointly with Louis Solomon and Hiroaki Terao [111].

I have received financial support from the National Science Foundation, the Wisconsin Alumni Research Foundation, the Mathematical Sciences Research Institute, Berkeley, and the Japan Society for the Promotion of Science. Parts of these notes were written at MSRI, Berkeley and at RIMS, Kyoto.

During the preparation of these lectures I visited several universities. I would like to thank my hosts for their hospitality: Eiichi Bannai in Columbus, Per Holm in Oslo, Haakon Waadeland in Trondheim, Michel Kervaire in Geneva, Rob Kirby and Emery Thomas in Berkeley, Kyoji Saito in Kyoto, and Mutsuo Oka in Tokyo.

Mike Falk's idea to organize this meeting gave the impetus to write these notes. He also helped me understand the work on minimal models. The presentation of the topological part owes a great deal to his PhD thesis [40], which was the first careful exposition of the foundational material. Arrangements are studied extensively by Soviet mathematicians. I am grateful to V. I. Arnold for references to this work. Louis Solomon and Hiroaki Terao taught me much of the contents of these notes and gave me permission to use material from our forthcoming book. I owe them special thanks.

Finally, I want to thank the participants of the conference in general, and Curtis Greene, Dick Randell, Tom Zaslavsky, and Sergey Yuzvinsky in particular, for their interest, enthusiasm, and help. The present version of the notes incorporates their suggestions for changes and corrections of the preliminary text distributed at the meeting.

Madison, October 23, 1988

1 Introduction

In this section we give the most important definitions, present some examples, and outline the contents of the notes.

Definitions and Examples

DEFINITION 1.1. *Let* **K** *be a field and let* $V_{\mathbf{K}}$ *be a vector space of dimension* ℓ. *A* **hyperplane** H *in* $V_{\mathbf{K}}$ *is a vector subspace of dimension* $(\ell - 1)$. *An* **arrangement** $\mathcal{A}_{\mathbf{K}} = (\mathcal{A}_{\mathbf{K}}, V_{\mathbf{K}})$ *is a finite set of hyperplanes in* $V_{\mathbf{K}}$.

The subscript **K** will be used only when we want to call attention to the field. In these notes we assume that **K** is the real or complex numbers. See [111] for interesting examples over finite fields. We call $\mathcal{A}$ an **ℓ-arrangement** when we want to emphasize the dimension of V. Let Φ_ℓ denote the empty ℓ-arrangement. Let V^* be the dual space of V, the space of linear forms on V. Let $S = S(V^*)$ be the symmetric algebra of V^*. Choose a basis $\{e_1, \ldots, e_\ell\}$ in V and let $\{x_1, \ldots, x_\ell\}$ be the dual basis in V^* so $x_i(e_j) = \delta_{i,j}$. We may identify $S(V^*)$ with the polynomial algebra $S = \mathbf{K}[x_1, \ldots, x_\ell]$. Each hyperplane $H \in \mathcal{A}$ is the kernel of a linear form α_H, defined up to a constant.

DEFINITION 1.2. *The product*

$$Q = Q(\mathcal{A}) = \prod_{H \in \mathcal{A}} \alpha_H$$

is called a **defining polynomial** *for* $\mathcal{A}$. *We agree that* $Q(\Phi_\ell) = 1$.

DEFINITION 1.3. *Arrangements in our sense are sometimes called* **central**. *If we view* V *as affine space and allow* $\mathcal{A}$ *to contain affine hyperplanes we call* $\mathcal{A}$ *an* **affine arrangement**. *If the intersection of all hyperplanes of an affine arrangement is not empty then coordinates may be chosen so that it becomes a central arrangement. If the intersection of all hyperplanes of an affine arrangement is empty we say that it is* **noncentral**.

REMARK 1.4. *Although many of the results in Sections 2,5,6,8 may be extended to affine arrangements, some important constructions in Sections 3,7,9 are only valid for central arrangements. We agree to use the term* **arrangement** *for an affine central arrangement. We attach the qualifier* **affine** *only when the arrangement may be noncentral, and on several occasions comment on the affine case. This may cause some confusion. These comments may be ignored at the expense of the loss of some generality by assuming that all arrangements are central.*

DEFINITION 1.5. *A* **projective arrangement** *is a finite set of projective hyperplanes in projective space.*

Since the complement of a hyperplane in projective space is affine space, a nonempty projective arrangement may be viewed as an affine arrangement. We shall not discuss projective arrangements separately. Sometimes we use $\mathcal{A}^*, \mathcal{B}^*, \ldots$ to denote affine arrangements constructed from the central arrangements $\mathcal{A}, \mathcal{B}, \ldots$. There is a close connection between central ℓ-arrangements and affine $(\ell - 1)$-arrangements. Let $\mathcal{A}$ be a central ℓ-arrangement with a distinguished hyperplane H. Choose coordinates so $H = \ker(x_\ell)$. Let Q be a defining polynomial for $\mathcal{A}$. To obtain an affine $(\ell - 1)$-arrangement consider the projective image of $\mathcal{A}$ where H is the hyperplane at infinity, and remove the image of H. We may identify the complement with affine $(\ell - 1)$-space and obtain an affine arrangement $\mathcal{A}^*$. Its defining polynomial $Q^* = Q(\mathcal{A}^*)$ is obtained by substituting 1 for x_ℓ in Q.

Conversely, given an affine $(\ell - 1)$-arrangement $\mathcal{A}^*$ with defining polynomial $Q^* = Q(\mathcal{A}^*)$, we may construct two central ℓ-arrangements from it: $\mathcal{A}_1$ has the same number of hyperplanes as $\mathcal{A}^*$ and its defining polynomial Q_1 is the polynomial Q^* homogenized; $\mathcal{A}_2$ has one more hyperplane and its defining polynomial is $Q_2 = Q_1 x_\ell$. The construction of the paragraph above applied to the central arrangement $\mathcal{A}_2$ recovers the affine arrangement $\mathcal{A}^*$. Although these constructions will reappear, it would overburden the notation if we introduced special names for them here.

First consider some real arrangements, $\mathbf{K} = \mathbf{R}$. If $\ell = 1$ then the only non-empty arrangement has the hyperplane $\{0\}$. For $\ell = 2, 3$ we agree to use x, y, z in place of x_1, x_2, x_3. A real 2-arrangement is a finite set of lines through the origin.

EXAMPLE 1.6. *Define* $\mathcal{A}_{\mathbf{R}}$ *by* $Q = xy(x + y)$. *It consists of three lines through the origin, see* Figure 1.

An affine real 2-arrangement is a finite set of lines in the plane.

EXAMPLE 1.7. *Define* $\mathcal{A}^*_{\mathbf{R}}$ *by* $Q^* = xy(x + y - 1)$. *It consists of three affine lines, see* Figure 2.

Real 3-arrangements are examples which display some of the intricacies of the general case.

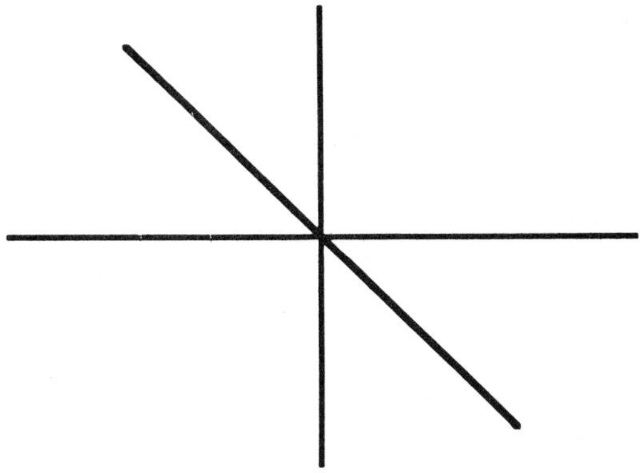

Figure 1: $Q = xy(x+y)$

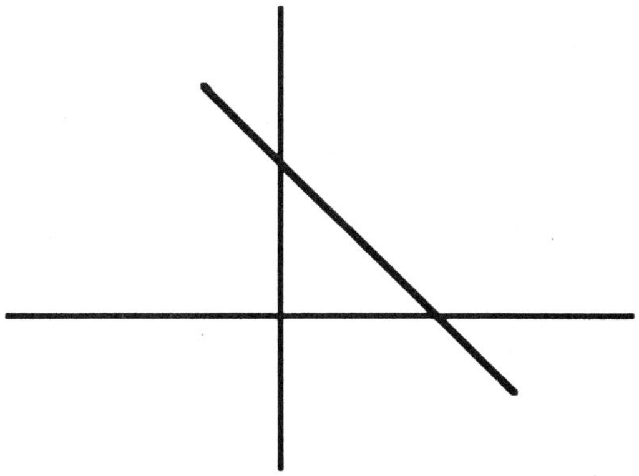

Figure 2: $Q^* = xy(x+y-1)$

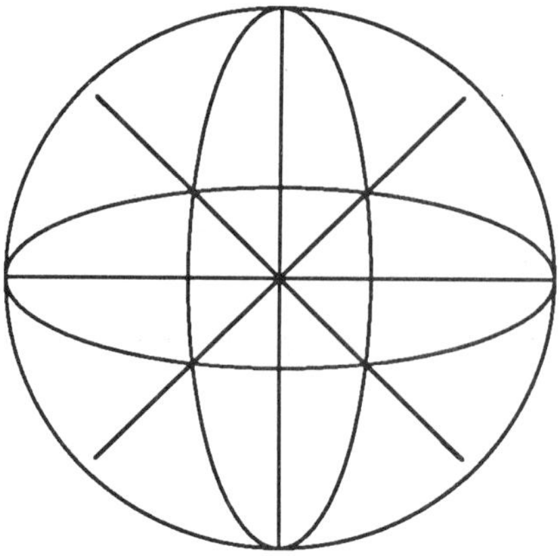

Figure 3: Projective image of the B_3-arrangement

EXAMPLE 1.8. *Let* $\mathbf{R}^3$ *have its usual basis. Consider the cube with vertices* $(\pm 1, \pm 1, \pm 1)$. *Its nine planes of symmetry form a 3-arrangement defined by*

$$Q = xyz(x - y)(x + y)(x - z)(x + z)(y - z)(y + z).$$

These nine planes intersect in lines, which are axes of rotational symmetry for the cube. The group of symmetries of the cube is the Coxeter group of type B_3. *We shall refer to this arrangement as the* B_3-**arrangement**.

We can visualize real 3-arrangements by passing to the projective plane. As usual, we think of the projective plane as the disk with identification of diametrically opposite points on the boundary. The picture we draw is therefore the intersection of the arrangement with the upper hemisphere of $S^2 \subset \mathbf{R}^3$. The corresponding picture for Example 1.8 is Figure 3.

The same idea may be conveyed by a slightly different picture, which is even easier to draw. If we assume that the line at infinity is in $\mathcal{A}$ then we may identify its complement in $\mathbf{RP}^2$ with $\mathbf{R}^2$ and draw the corresponding affine arrangement. *Here we must remember that parallel lines meet at infinity.* In Example 1.8 we let the plane $z = 0$ go to the line at infinity. If we substitute $z = 1$ in the remaining linear forms we get an affine 2-arrangement. In order to remember that the line at infinity is in our arrangement we draw a circle "far away" in Figure 4. It is easy to find and label the 13 lines of intersection in the 3-arrangement of Example 1.8 by finding the 13 points of intersection in either figure.

EXAMPLE 1.9. *Let* $\mathcal{A}_{\mathbf{R}}$ *be the* **Boolean** *arrangement defined by*

$$Q = x_1 x_2 \cdots x_\ell.$$

This is the arrangement of the coordinate hyperplanes in $\mathbf{R}^\ell$.

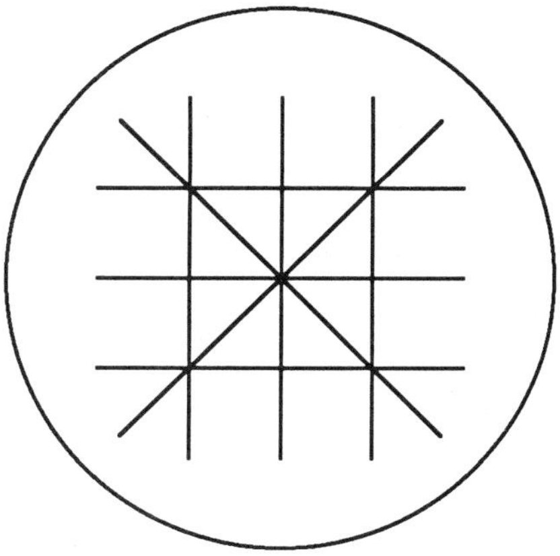

Figure 4: Another image of the B_3-arrangement

EXAMPLE 1.10. *For* $1 \le i < j \le \ell$ *let* $H_{i,j} = \ker(x_i - x_j)$. *Let* $\mathcal{A}_{\mathbf{R}}$ *be the arrangement defined by*

$$Q = \prod_{1 \le i < j \le \ell} (x_i - x_j).$$

This arrangement is familiar from the partition lattice and from complete graphs. It is also related to the braid group so we call it the **braid arrangement**.

DEFINITION 1.11. *Given an arrangement* $\mathcal{A}$ *we call the number of its hyperplanes* $|\mathcal{A}|$ *the* **cardinality** *of* $\mathcal{A}$.

In Example 1.9 we have $|\mathcal{A}| = \ell$. In Example 1.10 we have

$$|\mathcal{A}| = \ell(\ell - 1)/2.$$

It is clear from these examples that some of the complexity of $\mathcal{A}$ may be captured by knowledge of the intersection pattern of its hyperplanes.

DEFINITION 1.12. *Let* $L(\mathcal{A})$ *be the set of all intersections of elements of* $\mathcal{A}$. *We agree that* $L(\mathcal{A})$ *includes* V *as the empty intersection.*

We should remember that if $X \in L(\mathcal{A})$ then $X \subset V$. Strictly speaking these objects should have different names but it is always clear from the context which one is in consideration.

DEFINITION 1.13. *Let* $(\mathcal{A}, V)$ *be an arrangement. If* $\mathcal{B} \subset \mathcal{A}$ *is a subset then* $(\mathcal{B}, V)$ *is called a* **subarrangement**. *For* $X \in L(\mathcal{A})$ *define a subarrangement* $\mathcal{A}_X$ *of* $\mathcal{A}$ *by*

$$\mathcal{A}_X = \{H \in \mathcal{A} | X \subset H\},$$

and define an arrangement $(\mathcal{A}^X, X)$ *in* X *by*

$$\mathcal{A}^X = \{X \cap H | H \in \mathcal{A} - \mathcal{A}_X\}.$$

We call $\mathcal{A}^X$ *the* **restriction of** $\mathcal{A}$ **to** X.

The method of **deletion and restriction** is used frequently in inductive arguments. It refers to the following special case.

DEFINITION 1.14. *Let $\mathcal{A}$ be a nonempty arrangement and let $H \in \mathcal{A}$. Let $\mathcal{A}' = \mathcal{A} - \{H\}$ and let $\mathcal{A}'' = \mathcal{A}^H$. We call $(\mathcal{A}, \mathcal{A}', \mathcal{A}'')$ a* **triple** *of arrangements and H the* **distinguished** *hyperplane.*

Next we define the complexification of a real arrangement.

DEFINITION 1.15. *Let $(\mathcal{A}_{\mathbf{R}}, V_{\mathbf{R}})$ be an arrangement with defining polynomial $Q(\mathcal{A}_{\mathbf{R}})$. Its* **complexification** *is in $V = V_{\mathbf{R}} \otimes_{\mathbf{R}} \mathbf{C}$. It consists of the hyperplanes $\mathcal{A}_{\mathbf{C}} = \{H \otimes_{\mathbf{R}} \mathbf{C} | H \in \mathcal{A}_{\mathbf{R}}\}$. Thus $Q(\mathcal{A}_{\mathbf{C}}) = Q(\mathcal{A}_{\mathbf{R}})$.*

It is already quite difficult to visualize the complexification of Example 1.6. In real dimensions we have three 2-planes in 4-space which meet only at the origin. The complexification of the Boolean arrangement is the arrangement of the coordinate hyperplanes in $\mathbf{C}^\ell$.

The complexification of the braid arrangement occurs in the theory of configuration spaces and braids. Recall that a **braid** on ℓ strands may be viewed as the graph of the motion of ℓ distinct points in the complex line between times $t = 0$ and $t = 1$, subject to the condition that the points remain distinct throughout the motion. Thus we have a map $f : [0, 1] \to \mathbf{C}^\ell$ such that for each t the image point $(f_1(t), \dots, f_\ell(t))$ satisfies the condition $f_i(t) \neq f_j(t)$. The braid is **pure** if $f(0) = f(1)$. Thus a pure braid is the image of a circle in the complement of the hyperplanes $H_{i,j}$. The space $N = \bigcup_{1 \le i < j \le \ell} H_{i,j}$ is called the **superdiagonal** and its complement $M = V - N$ is the **pure braid space.**

DEFINITION 1.16. *Let $\mathcal{A}$ be an arrangement. Define*

$$N(\mathcal{A}) = \bigcup_{H \in \mathcal{A}} H,$$

and let

$$M(\mathcal{A}) = V - N(\mathcal{A}).$$

We call $M(\mathcal{A})$ the **complement** of the arrangement. It is the focal point of the topological study of complex arrangements. The variety $N(\mathcal{A})$ is a hypersurface with a very complicated singular set. In the present context the study of logarithmic vector fields and logarithmic differential forms on a hypersurface was initiated by Saito [124]. In the case of an arrangement we can pass from analytic to algebraic considerations.

DEFINITION 1.17. *Let $S = \mathbf{K}[x_1, \dots, x_\ell]$ be the algebra of polynomial functions on V. A $\mathbf{K}$-linear map*

$$\theta : S \to S$$

is a **derivation** *if for* $f, g \in S$

$$\theta(fg) = f\theta(g) + g\theta(f).$$

Let $\mathrm{Der}_{\mathbf{K}}(S)$ *be the* S*-module of derivations of* S.

DEFINITION 1.18. *Define an* S*-submodule of* $\mathrm{Der}_{\mathbf{K}}(S)$ *called the* **module of $\mathcal{A}$-derivations** *by*

$$D_S(\mathcal{A}) = \{\theta \in \mathrm{Der}_{\mathbf{K}}(S) | \theta(Q) \in QS\}.$$

DEFINITION 1.19. *The arrangement* $\mathcal{A}$ *is called* **free** *if the module* $D_S(\mathcal{A})$ *is a free* S*-module.*

Next we define a collection of arrangements with particularly nice properties. Let $GL(V)$ denote the general linear group of V.

DEFINITION 1.20. *An element* $s \in GL(V)$ *is a* **reflection** *if it has finite order and its fixed point set is a hyperplane* H_s. *We call* H_s *the* **reflecting hyperplane** *of* s. *A subgroup* $G \subset GL(V)$ *is called a* **reflection group** *if it is generated by reflections.*

DEFINITION 1.21. *Let* $G \subset GL(V)$ *be a finite reflection group. The set* $\mathcal{A} = \mathcal{A}(G)$ *of reflecting hyperplanes of* G *is called the* **reflection arrangement** *of* G.

Outline

Now that we have established the terminology we can give an outline of the contents of these notes. In Section 2 we study some basic combinatorial tools. The **intersection lattice** $L(\mathcal{A})$ is the most important combinatorial invariant of the arrangement $\mathcal{A}$. It is given a partial order by reverse inclusion and it is shown to be a geometric lattice. We define its Möbius function and study its properties. Next we define its Poincaré polynomial $\pi(\mathcal{A}, t)$, which is related to another combinatorial function called the characteristic polynomial. A fundamental technical tool is the method of **deletion and restriction**, which allows induction on the number of hyperplanes in the arrangement. It uses the triple $(\mathcal{A}, \mathcal{A}', \mathcal{A}'')$ of Definition 1.14.

We give an indication how the triple $(\mathcal{A}, \mathcal{A}', \mathcal{A}'')$ will be used. Assume that these arrangements are real and suppress the subscript $\mathbf{R}$. Since each real hyperplane disconnects space, the complement is a disjoint union of connected open subsets. Call each component a **chamber**. Let $\mathrm{Cham}(\mathcal{A})$ be the set of chambers. It is natural to ask how the arrangement $\mathcal{A}$ determines the number of its chambers $|\mathrm{Cham}(\mathcal{A})|$. It is clear from Example 1.10 that this is a nontrivial problem. The successful strategy uses the following observation:

PROPOSITION 1.22. *Let* $(\mathcal{A}, \mathcal{A}', \mathcal{A}'')$ *be a triple of real arrangements. Then*

$$|\mathrm{Cham}(\mathcal{A})| = |\mathrm{Cham}(\mathcal{A}')| + |\mathrm{Cham}(\mathcal{A}'')|.$$

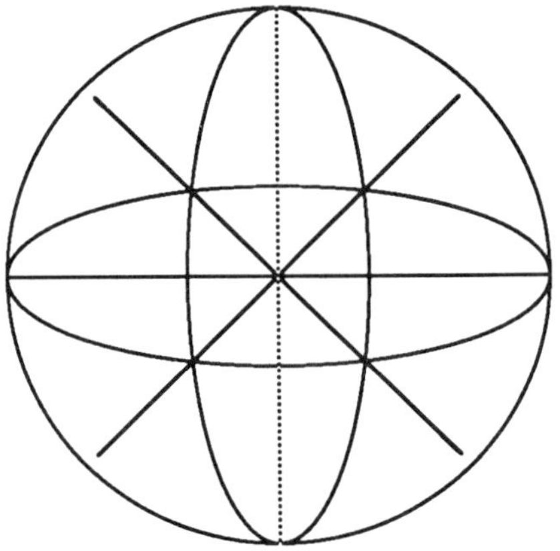

Figure 5: An illustration of chamber counting

Proof. Let P be the set of those chambers in $\mathrm{Cham}(\mathcal{A}')$ which intersect the distinguished hyperplane H. Let Q be the set of those chambers in $\mathrm{Cham}(\mathcal{A}')$ which do not intersect H. Evidently $|\mathrm{Cham}(\mathcal{A}')| = |P| + |Q|$. Now H separates each chamber of P into two chambers and leaves the chambers of Q uneffected. Thus $|\mathrm{Cham}(\mathcal{A})| = 2|P|+|Q|$. Finally, there is a bijection between P and $\mathrm{Cham}(\mathcal{A}'')$ given by $C \to C \cap H$. Thus $|\mathrm{Cham}(\mathcal{A}'')| = |P|$.

We illustrate this in Example 1.8. We must remember that the projective picture shows only half the chambers. Thus $|\mathrm{Cham}(\mathcal{A})| = 48$. If we let $H = \ker(x)$ be represented by a dotted line we see that $|P| = 8$, twice the four chambers adjacent to it in Figure 5, and $|Q| = 32$.

REMARK 1.23. *In order to show that there is a chamber counting function on $\mathcal{A}$, say $P(\mathcal{A})$ we need only check that P has two properties:*

(i) $P(\Phi_\ell) = 1$,

(ii) $P(\mathcal{A})$ *satisfies the recursion* $P(\mathcal{A}) = P(\mathcal{A}') + P(\mathcal{A}'')$.

We prove a theorem of Brylawski [22] about the characteristic polynomial under deletion and restriction. When this is applied to the Poincaré polynomial it gives the functional equation for a triple $(\mathcal{A}, \mathcal{A}', \mathcal{A}'')$:

$$\pi(\mathcal{A}, t) = \pi(\mathcal{A}', t) + t\pi(\mathcal{A}'', t).$$

Clearly $P(\mathcal{A}) = \pi(\mathcal{A}, 1)$ satisfies conditions (i) and (ii) above. Thus we obtain a theorem of Zaslavsky [157] for real arrangements:

$$|\mathrm{Cham}(\mathcal{A})| = \pi(\mathcal{A}, 1).$$

In Section 3 we introduce two graded algebras $A(\mathcal{A})$ and $B(\mathcal{A})$ constructed using only $L(\mathcal{A})$. Thus they are combinatorial invariants of $\mathcal{A}$. We state some of their basic properties. In particular their Poincaré polynomials equal

$\pi(\mathcal{A}, t)$. This gives an interpretation of the coefficients of $\pi(\mathcal{A}, t)$. The algebras are in fact isomorphic. In Section 4 we construct a simplicial complex $F(\mathcal{A})$ associated to $L(\mathcal{A})$ by Folkman [49]. We compute its homology groups and show that $F(\mathcal{A})$ has the homotopy type of a wedge of spheres. Section 4 contains another chain complex whose homology is naturally isomorphic to $B(\mathcal{A})$. We show how these constructions are related.

If $\mathcal{A}$ is a complex arrangement then the **complement of hyperplanes** $M(\mathcal{A})$ is its most important topological invariant. We study the topology of $M(\mathcal{A})$ next. In Section 5 we prove some elementary facts about the topology of $M(\mathcal{A})$ and discuss a few examples. We also give a description of fundamental work of Arnold, Brieskorn, Deligne and Hattori. In Section 6 we use a result of Brieskorn to show that the Poincaré polynomial of the complement, $P(M(\mathcal{A}), t) = \sum \text{rank} H^k(M(\mathcal{A}))t^k$, equals

$$P(M(\mathcal{A}), t) = \pi(\mathcal{A}, t).$$

Thus the coefficients of $\pi(\mathcal{A}, t)$ are also Betti numbers of the complement. This provides a topological interpretation of the triple $(\mathcal{A}, \mathcal{A}', \mathcal{A}'')$ as follows. There are split short exact sequences for all $k \geq 0$

$$0 \to H^k(M(\mathcal{A}')) \to H^k(M(\mathcal{A})) \to H^{k-1}(M(\mathcal{A}'')) \to 0.$$

Next we study the cohomology algebra of $M(\mathcal{A})$. We show that for each hyperplane H with linear form α_H the algebra $H^*(M(\mathcal{A}))$ has a generator $\omega_H = d\alpha_H/\alpha_H$ in degree 1, and the whole algebra is generated by these elements. We also discuss Falk's geometric linking and the results of Goresky and MacPherson on arrangements of subspaces of arbitrary codimension.

In Section 7 we associate to an arrangement $\mathcal{A}$ the algebra $R(\mathcal{A})$ generated by the elements ω_H. Note that this algebra is not a purely combinatorial object, since the linear forms enter the definition. The algebra $R(\mathcal{A})$ is defined for arrangements over any field. The main result of Section 7 is that there is an isomorphism of algebras $A(\mathcal{A}) \simeq R(\mathcal{A})$. This shows that $R(\mathcal{A})$ depends only on $L(\mathcal{A})$. The argument uses the fact that there is a short exact sequence of vector spaces

$$0 \to R(\mathcal{A}') \to R(\mathcal{A}) \to R(\mathcal{A}'') \to 0.$$

It follows by a dimension argument that for complex arrangements the map $R(\mathcal{A}) \to H^*(M(\mathcal{A}))$ which sends ω_H to its cohomology class is an algebra isomorphism. This is the topological interpretation of $A(\mathcal{A})$. In Section 8 we discuss recent work on the homotopy type of $M(\mathcal{A})$. We describe Salvetti's results on complexified real arrangements, the minimal model approach by Falk and Kohno, and Manin and Schechtman's discriminantal arrangements. The most important algebraic geometric invariant of $\mathcal{A}$ is the module $D_S(\mathcal{A})$. In Section 9 we study the algebraic properties of $D_S(\mathcal{A})$. We describe Terao's work on free arrangements, inductively free arrangements, and give several examples. Terao's factorization theorem, which asserts that if $\mathcal{A}$ is a free ℓ-arrangement then there exist non-negative integers $b_1, \ldots, b_\ell$ such that

$$\pi(\mathcal{A}, t) = (1 + b_1 t) \cdots (1 + b_\ell t)$$

is proved for inductively free arrangements. In Section 10 we assume the presence of a symmetry group. Suppose $(\mathcal{A}, V)$ is an arrangement and $G \subset GL(V)$ is a finite group such that $G(\mathcal{A}) \subset \mathcal{A}$. All our constructions may be done equivariantly. We obtain particularly nice results for the arrangements which arise as reflecting hyperplanes of unitary reflection groups. In particular every reflection arrangement is free. It follows from work of Brieskorn and Deligne that the complement of a complexified real reflection arrangement is a $K(\pi, 1)$ space. We conclude with an outline of the proof that the complements of the reflection arrangements of certain complex reflection groups, called Shephard groups, are also $K(\pi, 1)$ spaces.

There are several important topics that are closely related to arrangements but not discussed in these notes. Additional topological work is reviewed in the survey article by Falk and Randell [46]. In combinatorics there are several papers on various counting problems by Björner, Edelman and Ziegler [13], [15], [164], [165] and there are close connections with matroid theory. For a survey see Ziegler's thesis [163]. In algebraic geometry arrangements have been used by Hirzebruch [67], Sommese [132], and Hunt [68] to obtain varieties with interesting properties as branched covers of projective space with branch locus an arrangement. For a survey see [9]. Work of Gelfand and Zelevinski [53] on hypergeometric functions makes use of the algebra $A(\mathcal{A})$. Arrangements and group representations are combined in [100], in work of Lehrer [84], [85], and of Lehrer and Solomon [86].

The material is presented in the tradition of an introductory text. The exposition is detailed and contains proofs in the foundational parts. In the discussion of current research only the basic ideas and results are given, with an occasional hint of proof. The References include, in addition to those required in the text, an extensive bibliography of related work and preprints.

2 Combinatorics

In this section we present the necessary tools from combinatorics. We use Aigner's book [1] as a general reference for undefined terms.

The Lattice $L(\mathcal{A})$

DEFINITION 2.1. *Let $\mathcal{A}$ be an arrangement and let $L = L(\mathcal{A})$ be the set of all intersections of elements of $\mathcal{A}$. Define a* **partial order** *on L by:*

$$X \leq Y \Leftrightarrow Y \subseteq X.$$

Note that this is **reverse** inclusion. Thus V is the minimal element. Ordinary inclusion also gives a partial order but it does not have all the properties of a **geometric lattice** shown in Lemma 2.3.

DEFINITION 2.2. *Define a rank function on L by $r(X) = \operatorname{codim} X$. Thus $r(V) = 0$ and $r(H) = 1$ for $H \in \mathcal{A}$. Call H an* **atom** *of L. Define the* **join** *by $X \vee Y = X \cap Y$ and the* **meet** *by $X \wedge Y = \bigcap\{Z | Z \in L, X \cup Y \subset Z\}$.*

LEMMA 2.3. *Let $\mathcal{A}$ be an arrangement and let $L = L(\mathcal{A})$. Then:*
(i) *for every $X \in L$ all maximal linearly ordered subsets*

$$V = X_0 < X_1 < \cdots < X_p = X$$

have the same cardinality,
(ii) *every element of $L - \{V\}$ is a join of atoms,*
(iii) *for all $X, Y \in L$ the rank function satisfies*

$$r(X \wedge Y) + r(X \vee Y) \leq r(X) + r(Y).$$

Thus $L(\mathcal{A})$ is a geometric lattice.

Proof. Assertions (i) and (ii) are clear. To see (iii) recall that

$$\dim(X \cup Y) + \dim(X \cap Y) = \dim(X) + \dim(Y)$$

and $\dim(X \cup Y) \leq \dim(X \wedge Y)$.

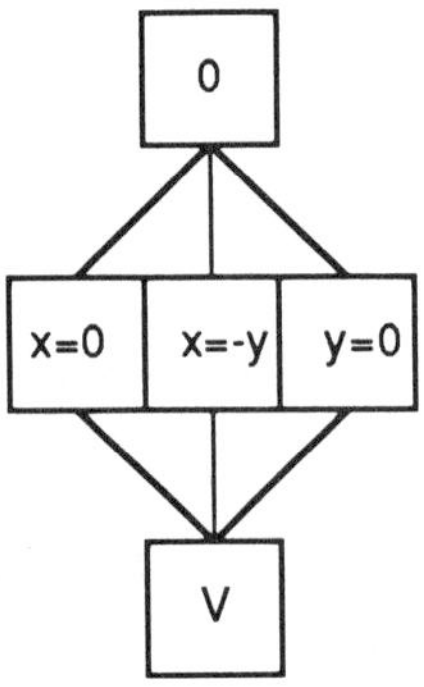

Figure 6: The Hasse diagram of $Q = xy(x + y)$

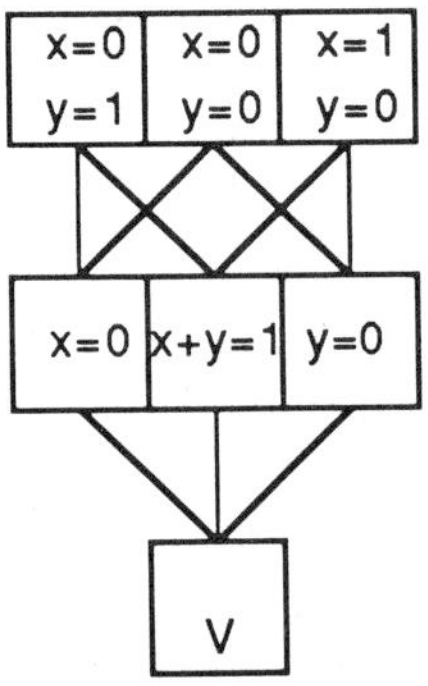

Figure 7: The Hasse diagram of $Q^* = xy(x + y - 1)$

Definition 2.4. *Let* $L_p(\mathcal{A}) = \{X \in L(\mathcal{A}) | r(X) = p\}$. *The* **Hasse diagram** *of* $L(\mathcal{A})$ *has vertices labeled by the elements of* $L(\mathcal{A})$ *and arranged on levels* $L_p, p \geq 0$. *Suppose* $X \in L_p$ *and* $Y \in L_{p+1}$. *An edge connects* X *with* Y *if* $X < Y$.

If $\mathcal{A}$ is an affine arrangement then $L(\mathcal{A})$ consists of the nonempty intersections of elements of $\mathcal{A}$. The partial order on $L(\mathcal{A})$ satisfies conditions (i) and (ii) of Lemma 2.3. The rank function is the same but not every pair in $L(\mathcal{A})$ has a join so $L(\mathcal{A})$ is not a lattice. It is called a **geometric semilattice**. The Hasse diagram is defined. If $\mathcal{A}$ is defined by a polynomial $Q(\mathcal{A})$ it is sometimes convenient to label elements of $L(\mathcal{A})$ by the equations they satisfy. The Hasse diagrams of Examples 1.6, 1.7, 1.8 appear in Figures 6, 7, 8.

Example 2.5. *The lattice of the Boolean arrangement.*

Let $H_i = \ker(x_i)$. Let $I = \{i_1, \ldots, i_p\}$ where $1 \leq i_1 < \cdots < i_p \leq \ell$. Let $H_I = H_{i_1} \cap \cdots \cap H_{i_p}$. The lattice L consists of the 2^ℓ subspaces H_I for all

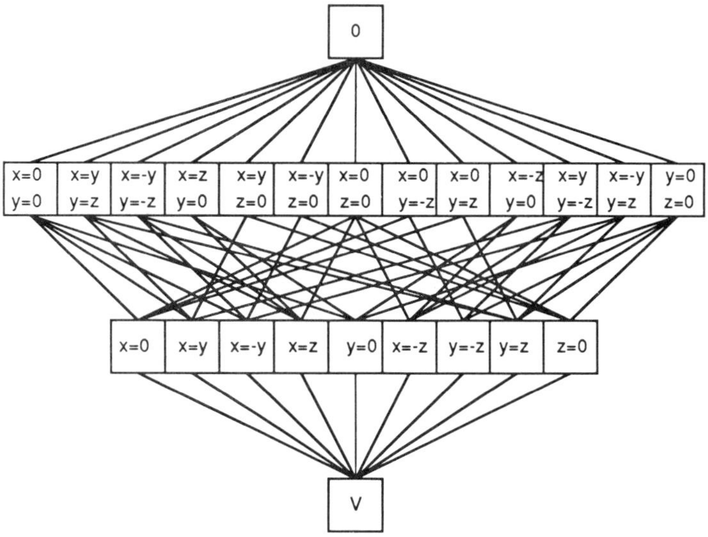

Figure 8: The Hasse diagram of the B_3-arrangement

subsets I.

DEFINITION 2.6. *Call the arrangements* $\mathcal{A}_1 = (\mathcal{A}_1, V_1)$ *and* $\mathcal{A}_2 = (\mathcal{A}_2, V_2)$ **lattice isomorphic** *if they have isomorphic lattices* $L(\mathcal{A}_1) \approx L(\mathcal{A}_2)$.

EXAMPLE 2.7. *The lattice of the braid arrangement is isomorphic to the* **partition lattice.**

Let $I = \{1, \ldots, \ell\}$. Let $\mathcal{P}(\ell)$ be the set of partitions of I. An element of $\mathcal{P}(\ell)$ is a collection $\Lambda = \{\Lambda_1, \ldots, \Lambda_r\}$ of nonempty pairwise disjoint subsets of I, called the blocks of Λ, whose union is I. There is a natural partial order on $\mathcal{P}(\ell)$ given by $\Lambda \leq \Gamma$ if Λ is finer than Γ. Thus blocks of Γ are unions of blocks of Λ. In order to find a lattice isomorphism from the braid lattice $L(\mathcal{A})$ to $\mathcal{P}(\ell)$ it is convenient to define $H_{i,i} = V$ for all i. Let $X \in L(\mathcal{A})$. Define a relation $\sim_X$ on I by $i \sim_X j$ if and only if $X \subseteq H_{i,j}$. Since $H_{i,i} = V$, $H_{i,j} = H_{j,i}$ and $H_{i,j} \cap H_{j,k} \subset H_{i,k}$, this is an equivalence relation. Let Λ_X be the partition of I defined by $\sim_X$. The map $\pi : L(\mathcal{A}) \to \mathcal{P}(\ell)$ given by $\pi(X) = \Lambda_X$ is a lattice isomorphism. It is injective because

$$X = \bigcap_{k=1}^{r} \left(\bigcap_{i,j \in \Lambda_k} H_{i,j} \right)$$

is determined by the blocks of Λ. It is surjective because given any partition $\Lambda = \{\Lambda_1, \ldots, \Lambda_r\}$ we may define X by the intersection above, and we get $\Lambda_X = \Lambda$. Note also that $X \leq Y$ if and only if every block of Λ_Y is a union of blocks of Λ_X.

DEFINITION 2.8. *Given a geometric lattice (semilattice) L and $X \in L$ let*

$$L_X = \{Z \in L | Z \leq X\},$$

$$L^X = \{Z \in L | Z \geq X\}.$$

LEMMA 2.9. *Let $\mathcal{A}$ be an arrangement and let $X \in L(\mathcal{A})$. Then*
(i) $L(\mathcal{A})_X = L(\mathcal{A}_X)$,
(ii) $L(\mathcal{A})^X = L(\mathcal{A}^X)$,
(iii) *if $Y \in L$ and $X \leq Y$ then*

$$L((\mathcal{A}_Y)^X) = L(\mathcal{A}_Y)^X = \{Z \in L | X \leq Z \leq Y\}.$$

DEFINITION 2.10. *Let $T(\mathcal{A}) = \bigcap_{H \in \mathcal{A}} H$ be the unique maximal element of $L(\mathcal{A})$, called the* **center** *of $\mathcal{A}$. The* **rank** *of $\mathcal{A}$ is $r(\mathcal{A}) = r(T(\mathcal{A}))$. Call the ℓ-arrangement $\mathcal{A}$* **essential** *if $r(\mathcal{A}) = \ell$.*

Thus $\mathcal{A}$ is essential if and only if it contains ℓ linearly independent hyperplanes, if and only if $T(\mathcal{A}) = \{0\}$. The braid arrangement is not essential, $T(\mathcal{A})$ is the line $x_1 = x_2 = \cdots = x_\ell$. All the other arrangements considered so far are essential.

DEFINITION 2.11. *Call $X, Y \in L(\mathcal{A})$* **incident** *if either $X < Y$ or $X > Y$. The* **incidence number** *$c^p_{q,m}$ is the number of elements in L_p incident with exactly m elements of L_q.*

In the B_3 arrangements the incidence numbers $c^2_{1,2} = 6$ and $c^2_{1,3} = 7$ represent the fact that there are six lines in two planes each, and seven lines in three planes each.

DEFINITION 2.12. *Let $(\mathcal{A}_1, V_1)$ and $(\mathcal{A}_2, V_2)$ be arrangements and let $V = V_1 \oplus V_2$. Define the* **direct product** *arrangement $(\mathcal{A}_1 \times \mathcal{A}_2, V)$ by*

$$\mathcal{A}_1 \times \mathcal{A}_2 = \{H_1 \oplus V_2 | H_1 \in \mathcal{A}_1\} \cup \{V_1 \oplus H_2 | H_2 \in \mathcal{A}_2\}.$$

There is a natural partial order on the set $L(\mathcal{A}_1) \times L(\mathcal{A}_2)$ of pairs (X_1, X_2) with $X_i \in L(\mathcal{A}_i)$:

$$(X_1, X_2) \leq (Y_1, Y_2) \Leftrightarrow X_1 \leq Y_1 \text{ and } X_2 \leq Y_2.$$

LEMMA 2.13. *Let $\mathcal{A}_1$, $\mathcal{A}_2$ be arrangements. There is a natural isomorphism of lattices*

$$\pi : L(\mathcal{A}_1) \times L(\mathcal{A}_2) \to L(\mathcal{A}_1 \times \mathcal{A}_2).$$

Proof. The map $\pi(X_1, X_2) = X_1 \oplus X_2$ provides the required isomorphism.

The Möbius Function

DEFINITION 2.14. *Let $\mathcal{A}$ be an arrangement and let $L = L(\mathcal{A})$. Define the* **Möbius function** $\mu_{\mathcal{A}} = \mu : L \times L \longrightarrow \mathbf{Z}$ *as follows:*

$$\begin{aligned} &\mu(X,X) = 1 && \text{if } X \in L, \\ &\sum_{X \le Z \le Y} \mu(X,Z) = 0 && \text{if } X,Y,Z \in L \text{ and } X < Y, \\ &\mu(X,Y) = 0 && \text{otherwise.} \end{aligned}$$

Note that for fixed X the values of $\mu(X,Y)$ may be computed recursively. It follows that if ν is any other function which satisfies the defining properties of μ then $\nu = \mu$. The definition of μ requires only a partially ordered set. Thus it is defined for affine arrangements.

There is a useful reformulation of $\mu(X,Y)$.

LEMMA 2.15. *Let $\mathcal{A}$ be an arrangement. Given $X, Y \in L$ such that $X \le Y$ let $S(X,Y)$ be the set of subarrangements $\mathcal{B} \subseteq \mathcal{A}$ such that $\mathcal{A}_X \subseteq \mathcal{B}$ and $T(\mathcal{B}) = Y$. Then*

$$\mu(X,Y) = \sum_{\mathcal{B} \in S(X,Y)} (-1)^{|\mathcal{B} - \mathcal{A}_X|}.$$

Proof. Let $\nu(X,Y)$ denote the right side of the expression. Note that

$$\bigcup_{X \le Z \le Y} S(X,Z) = \{\mathcal{B} \subseteq \mathcal{A} | \mathcal{A}_X \subseteq \mathcal{B} \subseteq \mathcal{A}_Y\}$$

where the union is disjoint. Thus

$$\sum_{X \le Z \le Y} \nu(X,Z) = \sum_{\mathcal{A}_X \subseteq \mathcal{B} \subseteq \mathcal{A}_Y} (-1)^{|\mathcal{B} - \mathcal{A}_X|} = \sum_{\mathcal{C}} (-1)^{|\mathcal{C}|}.$$

The last sum is over all subsets $\mathcal{C}$ of $\mathcal{A}_Y - \mathcal{A}_X$. If $X = Y$ the sum is 1. If $X < Y$ then $\mathcal{A}_X$ is a proper subset of $\mathcal{A}_Y$ so the sum is zero.

We shall prove that μ solves an inversion problem. First we need a lemma.

LEMMA 2.16. *Let $\mathcal{A}$ be an arrangement and let $L = L(\mathcal{A})$. Then*

$$\begin{aligned} &\mu(X,X) = 1 && \text{if } X \in L, \\ &\sum_{X \le Z \le Y} \mu(Z,Y) = 0 && \text{if } X,Y \in L \text{ and } X < Y. \end{aligned}$$

Proof. Write $L = \{X_1, \ldots, X_r\}$ where the numbering is chosen so that $X_i \le X_j$ implies $i \le j$. Let A be the $r \times r$ matrix with (i,j) entry $\mu(X_i, X_j)$. Let B be the $r \times r$ matrix with (i,j) entry 1 if $X_i \le X_j$ and 0 otherwise. Both A and B are upper unitriangular. It follows from the definition of μ that $AB = I_r$ is the identity matrix. Thus $BA = I_r$, which implies the assertions.

The next Proposition is the **Möbius inversion formula.**

PROPOSITION 2.17. *Let f, g be functions on $L(\mathcal{A})$ with values in an abelian group. Then*

$$g(Y) = \sum_{X \in L_Y} f(X) \Longleftrightarrow f(Y) = \sum_{X \in L_Y} \mu(X, Y) g(X),$$

$$g(X) = \sum_{Y \in L^X} f(Y) \Longleftrightarrow f(X) = \sum_{Y \in L^X} \mu(X, Y) g(Y).$$

Proof. Each of the four implications is based on an interchange of summation and the properties of μ given in the definition and in Lemma 2.16. We prove left to right implication in the first formula.

$$\begin{aligned} \sum_{Z \in L_Y} \mu(Z, Y) g(Z) &= \sum_{Z \in L_Y} \mu(Z, Y) \sum_{X \in L_Z} f(X) \\ &= \sum_{X \in L_Y} \left(\sum_{X \le Z \le Y} \mu(Z, Y) \right) f(X) \\ &= f(Y). \end{aligned}$$

The next result is due to Weisner [153].

LEMMA 2.18. *Let $\mathcal{A}$ be an arrangement and let $L = L(\mathcal{A})$.*
(i) *If $Y \in L$ and $Y \neq V$ then for all $Z \in L$*

$$\sum_{X \vee Y = Z} \mu(V, X) = 0.$$

(ii) *If $Y \in L$ and $Y \neq T = T(\mathcal{A})$ then for all $Z \in L$*

$$\sum_{X \wedge Y = Z} \mu(X, T) = 0.$$

Proof. We prove (i). The proof of (ii) is similar. Note that $X \vee Y = Z$ implies $r(Z) \geq r(Y)$. We argue by induction on $r(Z)$. If $Z = Y$ then the sum to be computed is $\sum_{X \leq Y} \mu(V, X) = 0$ since $Y \neq V$. If $Z > Y$ then

$$\begin{aligned} \sum_{X \vee Y = Z} \mu(V, X) &= \sum_{X \vee Y \le Z} \mu(V, X) - \sum_{X \vee Y < Z} \mu(V, X) \\ &= \sum_{X \le Z} \mu(V, X) - \sum_{W < Z} \left(\sum_{X \vee Y = W} \mu(V, X) \right). \end{aligned}$$

The first term is zero by the definition of μ. The second term is zero by induction.

In our applications the Möbius function with first variable fixed as V has special significance.

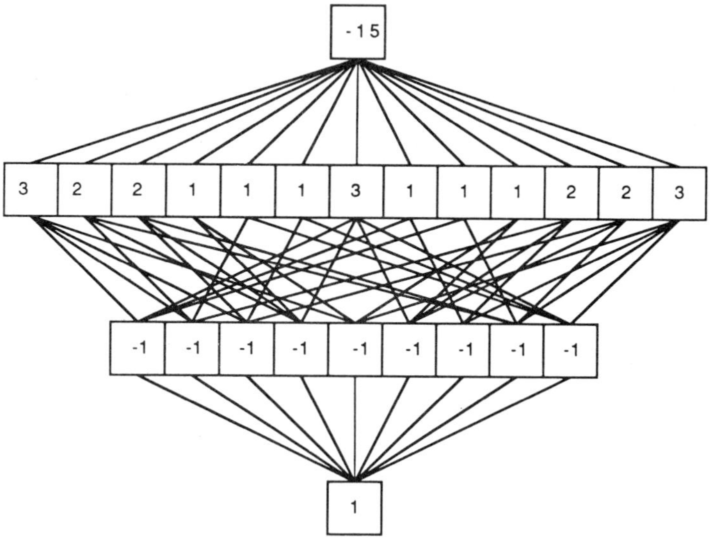

Figure 9: The values of $\mu(X)$ for the B_3-arrangement

Definition 2.19. *For $X \in L$ define $\mu(X) = \mu(V, X)$.*

Clearly $\mu(V) = 1$, $\mu(H) = -1$ for all $H \in L$ and if $r(X) = 2$ then $\mu(X) = |\mathcal{A}_X| - 1$. The values of $\mu(X)$ for all $X \in L$ are given for Example 1.8 in Figure 9. In general it is not possible to give a formula for $\mu(X)$ but we can compute all $\mu(X)$ in the Boolean lattice.

Proposition 2.20. *Define $\mathcal{A}$ by $Q = x_1 x_2 \cdots x_\ell$. Then for $X \in L$*

$$\mu(X) = (-1)^{r(X)}.$$

Proof. Define $\nu(X) = (-1)^{r(X)}$. It suffices to show that ν satisfies the defining properties of μ. Clearly $\nu(V) = (-1)^0 = 1$. If $X \neq V$ then $X = H_I$ for some I with $|I| = p = r(X) > 0$. If $V \leq Y \leq X$ then $Y = H_J$ where $J \subseteq I$ and

$$q = r(Y) = |J| \leq |I| = r(X) = p.$$

Thus

$$\sum_{Z \leq X} \nu(Z) = \sum_{J \subseteq I} (-1)^q = \sum_{q=0}^{p} (-1)^q \binom{p}{q} = 0.$$

Definition 2.21. *Let $\mu(\mathcal{A}) = \mu(T(\mathcal{A}))$.*

The next result is due to Rota [119].

Theorem 2.22. *Let $\mathcal{A}$ be an arrangement and let $L = L(\mathcal{A})$. If $X, Y \in L$ and $X \leq Y$ then $\mu(X, Y) \neq 0$ and* $\operatorname{sign}\mu(X, Y) = (-1)^{r(X) - r(Y)}$.

Proof. Since the Möbius function of $L(\mathcal{A}_Y)^X$ is the restriction of the Möbius function $\mu_{\mathcal{A}}$ we have $\mu(X,Y) = \mu((\mathcal{A}_Y)^X)$. Thus it suffices to show that $\mu(\mathcal{A}) \neq 0$ and sign $\mu(\mathcal{A}) = (-1)^{r(\mathcal{A})}$. We argue by induction on $r(\mathcal{A})$. The assertion is clear if $r(\mathcal{A}) = 0$. Suppose $r(\mathcal{A}) \geq 1$. Choose $H \in \mathcal{A}$ and apply Lemma 2.18.i with $Y = H$ and $Z = T(\mathcal{A})$. We get

$$\begin{aligned} 0 &= \mu(\mathcal{A}) + \sum_{X \in M} \mu(X) \\ &= \mu(\mathcal{A}) + \sum_{X \in M} \mu(\mathcal{A}_X) \end{aligned}$$

where M is the set of all $X \in L$ such that $X \neq T(\mathcal{A})$ but $X \vee H = T(\mathcal{A})$. If $X \in M$ then

$$\begin{aligned} r(\mathcal{A}) &= r(X \vee H) \\ &\leq r(X \vee H) + r(X \wedge H) \\ &\leq r(X) + r(H) \\ &= r(X) + 1. \end{aligned}$$

Thus $r(X) = r(\mathcal{A})-1$ and therefore $r(\mathcal{A}_X) = r(\mathcal{A})-1$. By induction $\mu(\mathcal{A}_X) \neq 0$ and sign $\mu(\mathcal{A}_X) = (-1)^{r(\mathcal{A}_X)} = (-1)^{r(\mathcal{A})-1}$. The assertion follows from the equation $\mu(\mathcal{A}) = -\sum_{X \in M} \mu(\mathcal{A}_X)$.

The Poincaré Polynomial of the Lattice

DEFINITION 2.23. *Let $\mathcal{A}$ be an arrangement with lattice L and Möbius function μ. Let t be an indeterminate. Define the* **Poincaré polynomial** *of $\mathcal{A}$ by*

$$\pi(\mathcal{A}, t) = \sum_{X \in L} \mu(X)(-t)^{r(X)}.$$

It follows from Theorem 2.22 that π has non-negative coefficients. This polynomial is also defined for affine arrangements, and most of the results of this section hold for affine arrangements. In some cases it is easy to compute the values of μ directly and obtain the Poincaré polynomial.

EXAMPLE 2.24. *If $\mathcal{A} = \Phi_\ell$ is the empty ℓ-arrangement then $\pi(\mathcal{A}, t) = 1$.*
The Poincaré polynomial in Example 1.6 *is:*

$$\pi(\mathcal{A}, t) = 1 + 3t + 2t^2 = (1+t)(1+2t).$$

The Poincaré polynomial in Example 1.7 *is:*

$$\pi(\mathcal{A}, t) = 1 + 3t + 3t^2.$$

The Poincaré polynomial in Example 1.8 *is:*

$$\pi(\mathcal{A}, t) = 1 + 9t + 23t^2 + 15t^3 = (1+t)(1+3t)(1+5t).$$

The Poincaré polynomial of the Boolean lattice is:

$$\pi(\mathcal{A}, t) = \sum_{p=0}^{\ell} \binom{\ell}{p} t^p = (1+t)^\ell.$$

These examples may give the false impression that the Poincaré polynomial of every central arrangement is a product of linear terms $(1+bt)$ where b is an integer. The reader is invited to check that if we homogenize Example 1.7 adding the line at infinity then $Q = xyz(x+y+z)$ defines an arrangement $\mathcal{A}$ with $\pi(\mathcal{A}, t) = 1 + 4t + 6t^2 + 3t^3 = (1+t)(1+3t+3t^2)$. It is not hard to show that $(1+t)$ divides $\pi(\mathcal{A}, t)$ for every central arrangement but more factors of the form $(1+bt)$ where b is an integer do not exist in general. The Poincaré polynomial is related to the following polynomial.

DEFINITION 2.25. *Define the* **characteristic polynomial** *of $\mathcal{A}$ by*

$$\chi(\mathcal{A}, t) = t^\ell \pi(\mathcal{A}, -t^{-1}) = \sum_{X \in L} \mu(X) t^{\dim(X)}.$$

Note that $\chi(\mathcal{A}, t)$ is a monic polynomial of degree ℓ. Our characteristic polynomial is slightly different from the usual definition of the characteristic polynomial of the lattice L. The definitions agree if $\mathcal{A}$ has rank ℓ. In some cases it is natural to compute the characteristic polynomial. Our first nontrivial computation obtains the characteristic polynomial of the braid arrangement. This computation has two interesting features: the combinatorial technique of proving an identity by expressing the cardinality of a set in two different ways, and use of Möbius inversion to find $\chi(\mathcal{A}, t)$ *without computing the individual values of* $\mu(X)$.

PROPOSITION 2.26. *Let $\mathcal{A}$ be the braid arrangement. Then*

$$\pi(\mathcal{A}, t) = (1+t)(1+2t)\cdots(1+(\ell-1)t).$$

Proof. We prove the equivalent formula:

$$\chi(\mathcal{A}, t) = t(t-1)(t-2)\cdots(t-(\ell-1)).$$

Let $I = \{1, \ldots, \ell\}$. Let W be a set with cardinality $|W| = w$. Let $M = W^I$ denote the set of all maps from I into W, so $|M| = w^\ell$. Each $\phi \in M$ determines an equivalence relation $\sim_\phi$ on I by $i \sim_\phi j$ if and only if $\phi(i) = \phi(j)$. Let Λ_ϕ be the corresponding partition. We classify the elements of M using the partitions Λ_ϕ. Given $X \in L(\mathcal{A})$ define subsets P_X and Q_X of M by

$$P_X = \{\phi \in M | \Lambda_\phi = \Lambda_X\}, \qquad Q_X = \{\phi \in M | \Lambda_\phi \geq \Lambda_X\}.$$

If $\Lambda_\phi \geq \Lambda_X$ then $\Lambda_\phi = \Lambda_Y$ for some $Y \geq X$ so that

$$Q_X = \bigcup_{Y \geq X} P_Y.$$

Thus by Möbius inversion

$$|P_Y| = \sum_{X \geq Y} \mu(Y,X)|Q_X|.$$

Let $B(X)$ be the set of blocks of Λ_X and let $b(X) = |B(X)|$. If $\phi \in Q_X$ then ϕ is constant on the blocks of Λ_X. Thus there is a bijection from Q_X to $W^{B(X)}$. In particular $|Q_X| = w^{b(X)}$.

Next note that $b(X) = \dim X$. We see this by choosing a basis for X consisting of vectors v^k defined by

$$v_i^k = \begin{cases} 1 & \text{if } i \in \Lambda_k, \\ 0 & \text{otherwise.} \end{cases}$$

In case $Y = V$ this gives

$$|P_V| = \sum_{X \in L} \mu(X) w^{\dim X}.$$

Since Λ_V is the partition where each block is a singleton, P_V is the set of one-to-one maps from I into W. Therefore we have

$$|P_V| = w(w-1)\cdots(w-(\ell-1)).$$

Since these formulas hold for all positive integers w, we are done.

The formula for $\mu(X,Y)$ obtained in Lemma 2.15 provides a useful expression for $\chi(\mathcal{A},t)$.

LEMMA 2.27. *Let $\mathcal{A}$ be an arrangement. Then*

$$\chi(\mathcal{A},t) = \sum_{\mathcal{B} \subseteq \mathcal{A}} (-1)^{|\mathcal{B}|} t^{\dim T(\mathcal{B})}.$$

Proof. Let $S(X) = S(V,X)$. From Lemma 2.15 we get

$$\chi(\mathcal{A},t) = \sum_{X \in L} \mu(X) t^{\dim X} = \sum_{X \in L} \left(\sum_{\mathcal{B} \in S(X)} (-1)^{|\mathcal{B}|} t^{\dim X} \right).$$

If $\mathcal{B} \in S(X)$ then $T(\mathcal{B}) = X$ so $\dim T(\mathcal{B}) = \dim X$. The result follows since every subset $\mathcal{B}$ of $\mathcal{A}$ occurs in a unique $S(X)$.

We are now prepared for the main result of this section, the **Deletion-Restriction Theorem**. This result was proved by Brylawski [22] for central arrangements and by Zaslavsky [157] in general.

THEOREM 2.28. *Let $(\mathcal{A}, \mathcal{A}', \mathcal{A}'')$ be a triple of arrangements. Then*

$$\chi(\mathcal{A},t) = \chi(\mathcal{A}',t) - \chi(\mathcal{A}'',t).$$

Proof. We use the formula in Lemma 2.27. Separate the sum over $\mathcal{B} \subseteq \mathcal{A}$ into two sums: R' and R''. Here R' is the sum over those $\mathcal{B}$ which do not contain H, and R'' is the sum over those $\mathcal{B}$ which contain H. It follows from Lemma 2.27 with $\mathcal{A}'$ in place of $\mathcal{A}$ that

$$R' = \chi(\mathcal{A}', t).$$

In order to compute R'' recall the definition of $S(X, Y)$ from Lemma 2.15. Since $H \in \mathcal{B}$, $\mathcal{A}_H \subseteq \mathcal{B}$. Thus if $T(\mathcal{B}) = Y$ then $\mathcal{B} \in S(H, Y)$. Let $L'' = L(\mathcal{A}'')$. Then

$$\begin{aligned} R'' &= \sum_{H \in \mathcal{B} \subseteq \mathcal{A}} (-1)^{|\mathcal{B}|} t^{\dim T(\mathcal{B})} \\ &= \sum_{Y \in L''} \sum_{\mathcal{B} \in S(H,Y)} (-1)^{|\mathcal{B}|} t^{\dim Y} \\ &= -\sum_{Y \in L''} \sum_{\mathcal{B} \in S(H,Y)} (-1)^{|\mathcal{B} - \mathcal{A}_H|} t^{\dim Y} \\ &= -\sum_{Y \in L''} \mu(H, Y) t^{\dim Y} \\ &= -\chi(\mathcal{A}'', t). \end{aligned}$$

The last equality follows from Lemma 2.15 and the fact that the Möbius function $\mu_{\mathcal{A}''}$ of L'' is the restriction of μ to L'' so $\mu_{\mathcal{A}''}(Y) = \mu(H, Y)$.

COROLLARY 2.29. *Let $(\mathcal{A}, \mathcal{A}', \mathcal{A}'')$ be a triple of arrangements. Then*

$$\pi(\mathcal{A}, t) = \pi(\mathcal{A}', t) + t\pi(\mathcal{A}'', t).$$

DEFINITION 2.30. *Let $(\mathcal{A}, \mathcal{A}', \mathcal{A}'')$ be a triple with respect to the hyperplane $H \in \mathcal{A}$. Call H a* **separator** *if $T(\mathcal{A}) \notin L(\mathcal{A}')$.*

COROLLARY 2.31. *Let $(\mathcal{A}, \mathcal{A}', \mathcal{A}'')$ be a triple with respect to $H \in \mathcal{A}$.*

(i) *If H is a separator then*

$$\mu(\mathcal{A}) = -\mu(\mathcal{A}'')$$

and hence

$$|\mu(\mathcal{A})| = |\mu(\mathcal{A}'')|.$$

(ii) *If H is not a separator then*

$$\mu(\mathcal{A}) = \mu(\mathcal{A}') - \mu(\mathcal{A}'')$$

and

$$|\mu(\mathcal{A})| = |\mu(\mathcal{A}')| + |\mu(\mathcal{A}'')|.$$

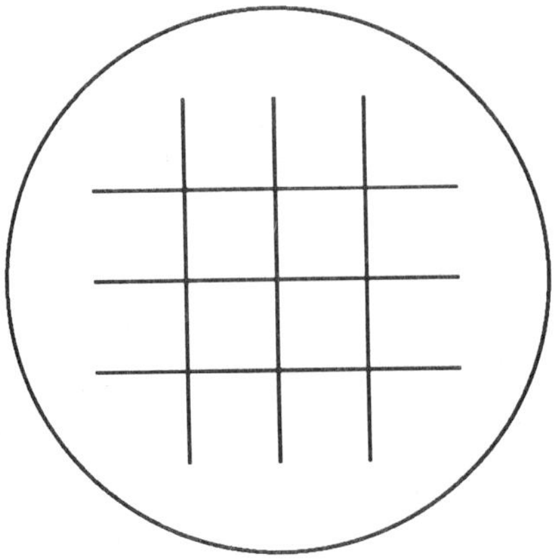

Figure 10: $Q(\mathcal{A}_1) = xyz(x-z)(x+z)(y-z)(y+z)$

Proof. It follows from Theorem 2.22 that $\pi(\mathcal{A}, t)$ has leading term

$$(-1)^{r(\mathcal{A})}\mu(\mathcal{A})t^{r(\mathcal{A})}.$$

The conclusion follows by comparing coefficients of the leading terms on both sides of the equation in Corollary 2.29. If H is a separator then $r(\mathcal{A}') < r(\mathcal{A})$ and there is no contribution from $\pi(\mathcal{A}', t)$.

The Poincaré polynomial of an arrangement will appear repeatedly in these notes. It will be shown to equal the Poincaré polynomial of the graded algebras which we are going to associate with $\mathcal{A}$. It is also the Poincaré polynomial of the complement $M(\mathcal{A})$ for a complex arrangement. Here we prove that the Poincaré polynomial is the chamber counting function for a real arrangement. The complement $M(\mathcal{A})$ is a disjoint union of chambers

$$M(\mathcal{A}) = \bigcup_{C \in \mathrm{Cham}(\mathcal{A})} C.$$

Zaslavsky [157] showed that the number of chambers is determined by the Poincaré polynomial as follows.

THEOREM 2.32. *Let $\mathcal{A}_{\mathbf{R}}$ be a real arrangement. Then*

$$|\mathrm{Cham}(\mathcal{A}_{\mathbf{R}})| = \pi(\mathcal{A}_{\mathbf{R}}, 1).$$

Proof. We check the properties required in Remark 1.23: (i) follows from $\pi(\Phi_\ell, t) = 1$, and (ii) is a consequence of Corollary 2.29.

EXAMPLE 2.33. *Lattice isomorphic arrangements have the same Poincaré polynomial. The converse is false. Consider the arrangements:*

$$Q(\mathcal{A}_1) = xyz(x-z)(x+z)(y-z)(y+z),$$

$$Q(\mathcal{A}_2) = xyz(x+y+z)(x+y-z)(x-y+z)(x-y-z).$$

Figure 11: $Q(\mathcal{A}_2) = xyz(x+y+z)(x+y-z)(x-y+z)(x-y-z)$

The reader should check that

$$\pi(\mathcal{A}_1, t) = \pi(\mathcal{A}_2, t) = (1+t)(1+3t)(1+3t)$$

but their lattices are not isomorphic. For example $\mathcal{A}_1$ has two lines which are contained in four hyperplanes, $c_{1,4}^2(\mathcal{A}_1) = 2$. These appear in Figure 10 as the two common points on the line at infinity of the two sets of three parallel lines. Figure 11 shows that $\mathcal{A}_2$ has no such lines, $c_{1,4}^2(\mathcal{A}_2) = 0$.

3 Combinatorial Algebras

In the first part of this section we associate to the arrangement $\mathcal{A}$ a graded anticommutative algebra $A(\mathcal{A})$ which depends only on the lattice $L(\mathcal{A})$. Thus it is a combinatorial invariant of $\mathcal{A}$. We prove that its Poincaré polynomial is $\pi(\mathcal{A}, t)$. Next we define a second graded algebra $B(\mathcal{A})$ whose elements are certain **K**-linear combinations of ordered subsets of $L(\mathcal{A})$, with multiplication defined by using a shuffle product and show that these algebras are isomorphic. The algebra $A(\mathcal{A})$ was introduced in [100], where it was used to prove that for a complex arrangement $A(\mathcal{A})$ is isomorphic as a graded algebra to the cohomology of the complement $M(\mathcal{A})$. We show this in Section 6. The algebra $A(\mathcal{A})$ has since been used by Gelfand and Zelevinski [53] in their work on hypergeometric functions. The algebra $B(\mathcal{A})$ was also defined in [100], where it was shown that $B(\mathcal{A})$ is isomorphic as vector space to a truncated homology defined on $L(\mathcal{A})$ by Baclawski [6]. We show this in Section 4.

The Algebra $A(\mathcal{A})$

DEFINITION 3.1. *Let $\mathcal{A}$ be an arrangement over* **K**. *Let*

$$E_1 = \bigoplus_{H \in \mathcal{A}} \mathbf{K} e_H$$

and let

$$E = E(\mathcal{A}) = \Lambda(E_1)$$

be the exterior algebra of E_1.

Note that E_1 has a **K**-basis consisting of elements e_H in one-to-one correspondence with the hyperplanes of $\mathcal{A}$. Write $uv = u \wedge v$ and note that $e_H^2 = 0$, $e_H e_K = -e_K e_H$ for $H, K \in \mathcal{A}$. The algebra E is graded. If $|\mathcal{A}| = n$ then

$$E = \bigoplus_{p=0}^{n} E_p$$

where $E_0 = \mathbf{K}$, E_1 agrees with its earlier definition and E_p is spanned over **K** by all $e_{H_1} \cdots e_{H_p}$ with $H_k \in \mathcal{A}$.

DEFINITION 3.2. *Define a* **K**-*linear map* $\partial_E = \partial : E \to E$ *by* $\partial 1 = 0$, $\partial e_H = 1$ *and for* $p \geq 2$

$$\partial(e_{H_1} \cdots e_{H_p}) = \sum_{k=1}^{p} (-1)^{k-1} e_{H_1} \cdots \widehat{e_{H_k}} \cdots e_{H_p}$$

for all $H_1, \ldots, H_p \in \mathcal{A}$.

DEFINITION 3.3. *Given a p-tuple of hyperplanes* $S = (H_1, \ldots, H_p)$ *write* $|S| = p$,

$$e_S = e_{H_1} \cdots e_{H_p} \in E,$$

and

$$\bigcap S = H_1 \cap \cdots \cap H_p \in L.$$

If $p = 0$ *we agree that* $S = (\ )$ *is the empty tuple,* $e_S = 1$ *and* $\bigcap S = V$.

It is convenient to introduce some more notation. If $S = (H_1, \ldots, H_p)$ we say that $H_i \in S$. If T is a subsequence of S we write $T \subset S$. If $T = (K_1, \ldots, K_q)$ we write $(S, T) = (H_1, \ldots, H_p, K_1, \ldots, K_q)$. Thus $e_{(S,T)} = e_S e_T$ and in particular for $H \in \mathcal{A}$ we have $e_{(H,S)} = e_H e_S$. Let $\mathbf{S}_p$ denote the set of all p-tuples $(H_1, \ldots, H_p)$ and let $\mathbf{S} = \bigcup_{p \geq 0} \mathbf{S}_p$.

LEMMA 3.4. *The map* $\partial : E \to E$ *satisfies:*
(i) $\partial^2 = 0$,
(ii) *If* $u \in E_p$ *and* $v \in E$ *then*

$$\partial(uv) = (\partial u)v + (-1)^p u(\partial v).$$

Proof. Part (i) is the standard boundary formula. It suffices to check (ii) for $u = e_S$ and $v = e_T$ for $S, T \in \mathbf{S}$, where it follows by direct computation.

Note that this lemma has nothing to do with arrangements. It states two familiar properties of the exterior algebra. Since the map ∂ is homogeneous of degree -1 we see from (i) that (E, ∂) is a chain complex. Part (ii) says that ∂ is a derivation of the exterior algebra. It may be characterized as the unique derivation of E with $\partial e_H = 1$. Since the rank function on L is codimension, it is clear that if $|S| = p$ then $r(\bigcap S) \leq p$.

DEFINITION 3.5. *Let* $S = (H_1, \ldots, H_p)$. *Call* S **independent** *if* $r(\bigcap S) = p$ *and* **dependent** *if* $r(\bigcap S) < p$.

The terminology has geometric significance. The tuple S is independent if the corresponding linear forms $\alpha_1, \ldots, \alpha_p$ are linearly independent. Equivalently, the hyperplanes of S are in general position.

DEFINITION 3.6. *Let* $\mathcal{A}$ *be an arrangement. Let* $I = I(\mathcal{A})$ *be the ideal of* E *generated by* ∂e_S *for all dependent* $S \in \mathbf{S}$.

Since I is generated by homogeneous elements, it is a graded ideal. Let $I_p = I \cap E_p$. Then

$$I = \bigoplus_{p=0}^{n} I_p.$$

Definition 3.7. *Let $\mathcal{A}$ be an arrangement. Let $A = A(\mathcal{A}) = E/I$. Let $\varphi : E \to A$ be the natural homomorphism and let $A_p = \varphi(E_p)$. If $H \in \mathcal{A}$ let $a_H = \varphi(e_H)$ and if $S \in \mathbf{S}$ let $a_S = \varphi(e_S)$.*

Lemma 3.8. *If $S \in \mathbf{S}$ and $H \in S$ then $e_S = e_H(\partial e_S)$. If S is dependent then $e_S \in I$.*

Proof. If H is in S then $e_H e_S = 0$ so that $0 = \partial(e_H e_S) = e_S - e_H(\partial e_S)$. The second assertion follows from the first.

Since both E and I are graded, A is a graded anticommutative algebra. Since the elements of $\mathbf{S}_1$ are independent we have $I_0 = 0$ and hence $A_0 = \mathbf{K}$. The only dependent elements of $\mathbf{S}_2$ are of the form $S = (H, H)$. Since $e_S = e_H^2 = 0$ we have $I_1 = 0$. Thus the elements a_H are linearly independent over $\mathbf{K}$ and $A_1 = \bigoplus_{H \in \mathcal{A}} a_H$. If $p > \ell$ then every element of $\mathbf{S}_p$ is dependent and it follows from Lemma 3.8 that $A_p = 0$. Thus

$$A = \bigoplus_{p=0}^{\ell} A_p.$$

Example 3.9. *Suppose $\ell = 2$ and $\mathcal{A} = \{H_1, ..., H_n\}$. Write $a_k = a_{H_k}$. Then*

$$A(\mathcal{A}) = \mathbf{K} \oplus \bigoplus_{p=1}^{n} \mathbf{K} a_p \oplus \bigoplus_{k=1}^{n-1} \mathbf{K} a_k a_n.$$

We have computed A_0, A_1 and we know that $A_p = 0$ for $p > 2$. It remains to compute A_2. Since $\dim V = 2$, (H_i, H_j, H_k) is dependent for all (i, j, k). Thus I_2 contains the element

$$\partial(e_i e_j e_k) = e_j e_k - e_i e_k + e_i e_j = e_i e_j + e_j e_k + e_k e_i.$$

Thus A_2 is spanned by $a_p a_q$ subject to the relations

$$a_i a_j + a_j a_k + a_k a_i = 0$$

for all (i, j, k). This shows that A_2 is spanned by $a_k a_n$ for $1 \le k < n$. It remains to show that the sum is direct. Suppose $\sum_{k=1}^{n-1} c_k a_k a_n = 0$ with $c_k \in \mathbf{K}$. Then $\sum_{k=1}^{n-1} c_k e_k e_n \in I_2$. Recall that I_2 is spanned by the elements $\partial(e_i e_j e_k)$. Since $\partial\partial = 0$ we have $\partial I_2 = 0$ and hence

$$\partial\left(\sum_{k=1}^{n-1} c_k e_k e_n\right) = \sum_{k=1}^{n-1} c_k (e_n - e_k) = 0.$$

Since $e_1, \dots, e_n$ are linearly independent over $\mathbf{K}$ we conclude that $c_k = 0$ for all k.

EXAMPLE 3.10. *If $\mathcal{A}$ is the Boolean arrangement then $S = (H_1, \dots, H_p)$ is independent if and only if $H_1, \dots, H_p$ are distinct hyperplanes. Hence if S is dependent then $e_S = 0$. Thus $I = 0$ and $A = E$.*

LEMMA 3.11. $\partial_E I \subset I$.

Proof. Recall that I is a $\mathbf{K}$ linear combination of elements of the form $e_T \partial e_S$ where $T, S \in \mathbf{S}$ and S is dependent. We have

$$\partial(e_T \partial e_S) = (\partial e_T)(\partial e_S) \pm e_T(\partial^2 e_S) = (\partial e_T)(\partial e_S) \in I.$$

DEFINITION 3.12. *Since $\partial_E I \subset I$ we may define $\partial_A : A \to A$ by $\partial_A \varphi u = \varphi \partial_E u$ for $u \in E$.*

LEMMA 3.13. *The map $\partial_A : A \to A$ satisfies*

(i) $\partial_A^2 = 0$,

(ii) *If $a \in A_p$ and $b \in A$ then*

$$\partial_A(ab) = (\partial_A a)b + (-1)^p a(\partial_A b).$$

(iii) *If $\mathcal{A}$ is not empty then the chain complex (A, ∂_A) is acyclic.*

Proof. Parts (i) and (ii) follow from the corresponding facts for ∂_E. Since ∂_A is homogeneous of degree -1, (A, ∂_A) is a chain complex. It follows from (i) that $\operatorname{im}\partial_A \subset \ker \partial_A$. To prove that the complex is acyclic we must show the reverse inclusion. Since $\mathcal{A}$ is not empty we may choose $H \in \mathcal{A}$. Let $v = e_H$. Then $\partial_E v = 1$. Let $b = \varphi v$ and let $a \in A$. Choose $u \in E$ with $\varphi u = a$. Then

$$\partial_E(vu) = (\partial_E v)u - v(\partial_E u) = u - v(\partial_E u).$$

Since $\varphi \partial_E = \partial_A \varphi$ and φ is a $\mathbf{K}$-algebra homomorphism, applying φ to the first and last terms gives

$$a = \partial_A(ba) + b\partial_A a$$

for all $a \in A$. Thus $\operatorname{im}\partial_A \supset \ker \partial_A$.

PROPOSITION 3.14. *Suppose $\mathcal{A}$ is not empty. Then*

$$\sum_{p \geq 0} (-1)^p \dim A_p = 0.$$

Proof. Since the chain complex (A, ∂_A) is acyclic, its Euler characteristic is zero.

Next we state some facts about I and A. The proofs are in [100], [109], [111]. Let

$$\mathbf{S}_X = \{S \in \mathbf{S} | X = \bigcap S\}.$$

Let $E_X = \sum_{S \in \mathbf{S}_X} \mathbf{K} e_S$ and let $I_X = I \cap E_X$.

PROPOSITION 3.15. $I = \bigoplus_{X \in L} I_X$.

DEFINITION 3.16. *If* $X \in L$ *let* $A_X = \varphi(E_X)$.

THEOREM 3.17. *Let* $\mathcal{A}$ *be an arrangement and let* $A = A(\mathcal{A})$. *Then*

$$A = \bigoplus_{X \in L} A_X.$$

COROLLARY 3.18. $A_p = \bigoplus_{X \in L_p} A_X$.

LEMMA 3.19. *If* $X \in L(\mathcal{A})$ *then* $\dim A_X(\mathcal{A}_X) = \dim A_X(\mathcal{A})$.

PROPOSITION 3.20. *If* $X \in L(\mathcal{A})$ *then* $\dim A_X = (-1)^{r(X)}\mu(X)$.

Proof. We argue by induction on $r(\mathcal{A})$. If $r(\mathcal{A}) = 0$ then $\mathcal{A}$ is empty and $A = \mathbf{K}$ so the assertion holds. If $r(\mathcal{A}) = 1$ then $\mathcal{A} = \{H\}$ so $A = \mathbf{K} \oplus \mathbf{K}a_H$ and the assertion is clear. Suppose $r(\mathcal{A}) \geq 2$. For $X \neq T(\mathcal{A})$ we have $r(\mathcal{A}_X) < r(\mathcal{A})$. Let μ_X be the Möbius function of $L(\mathcal{A}_X)$. By induction $\dim A_X(\mathcal{A}_X) = (-1)^{r(X)}\mu_X(X) = (-1)^{r(X)}\mu(X)$. Thus $\dim A_X(\mathcal{A}) = (-1)^{r(X)}\mu(X)$ for all $X \neq T(\mathcal{A})$. Let $T = T(\mathcal{A})$. From 3.14 and 3.17 we get

$$\begin{aligned} 0 &= \sum_{p \geq 0} (-1)^p \dim A_p(\mathcal{A}) \\ &= \sum_{X \in L(\mathcal{A})} (-1)^{r(X)} \dim A_X(\mathcal{A}) \\ &= (-1)^{r(T)} \dim A_T(\mathcal{A}) + \sum_{X \neq T} \mu(X). \end{aligned}$$

Since $\sum_{X \in L(\mathcal{A})} \mu(X) = 0$ this shows that $\mu(T) = (-1)^{r(T)} \dim A_T(\mathcal{A})$ as required.

THEOREM 3.21. *Let* $\mathcal{A}$ *be an arrangement and let* $A(\mathcal{A})$ *be the associated algebra. Let* $P(A(\mathcal{A}), t)$ *be the Poincaré polynomial of the graded* $\mathbf{K}$*-algebra* $A(\mathcal{A})$. *Then*

$$P(A(\mathcal{A}), t) = \pi(\mathcal{A}, t).$$

THEOREM 3.22. *Let* $\mathcal{A}$ *be an arrangement. Let* $H_0 \in \mathcal{A}$ *and let* $(\mathcal{A}, \mathcal{A}', \mathcal{A}'')$ *be the corresponding triple. Let* $i : A(\mathcal{A}') \to A(\mathcal{A})$ *be the natural homomorphism and let* $j : A(\mathcal{A}) \to A(\mathcal{A}'')$ *be the* $\mathbf{K}$*-linear map defined by*

$$j(a_{H_1} \cdots a_{H_p}) = 0,$$

$$j(a_{H_0} a_{H_1} \cdots a_{H_p}) = a_{H_0 \cap H_1} \cdots a_{H_0 \cap H_p}$$

for $(H_1, \ldots, H_p) \in \mathbf{S}(\mathcal{A}')$. *Then the sequence*

$$0 \to A(\mathcal{A}') \xrightarrow{i} A(\mathcal{A}) \xrightarrow{j} A(\mathcal{A}'') \to 0$$

is exact.

The Algebra $B(\mathcal{A})$

DEFINITION 3.23. *Let $\mathcal{A}$ be an arrangement with lattice $L = L(\mathcal{A})$. For $p \geq 0$ define vector spaces $\mathcal{T}_p$ as follows: $\mathcal{T}_0 = \mathbf{K}$ and for $p > 0$ $\mathcal{T}_p$ has a basis consisting of all p-tuples $(X_1, \ldots, X_p)$ where $X_i \in L - \{V\}$. Let*

$$\mathcal{T} = \bigoplus_{p \geq 0} \mathcal{T}_p.$$

Let $\mathrm{Sym}(p)$ be the symmetric group on the letters $1, \ldots, p$. If $\pi \in \mathrm{Sym}(p)$ and $u = (X_1, \ldots, X_p)$ let $\pi u = (X_{\pi^{-1}1}, \ldots, X_{\pi^{-1}p})$. This makes $\mathcal{T}_p$ a $\mathrm{Sym}(p)$-module.

DEFINITION 3.24. *Define a product $\mathcal{T} \times \mathcal{T} \to \mathcal{T}$, written $*$, as follows. If $u = (X_1, \ldots, X_p)$ and $v = (Y_1, \ldots, Y_q)$ let*

$$w = (Z_1, \ldots, Z_{p+q}) = (X_1, \ldots, X_p, Y_1, \ldots, Y_q).$$

Define

$$u * v = \sum \operatorname{sign} \pi (\pi w)$$

where the sum is over all (p, q)-shuffles π of $1, \ldots, p + q$.

Recall [88, p. 243] that a (p, q)-shuffle of $1, \ldots, p + q$ is a permutation $\pi \in \mathrm{Sym}(p + q)$ such that $\pi i < \pi j$ whenever $i < j \leq p$ or $p < i < j$. This makes $\mathcal{T}$ into an associative graded anticommutative $\mathbf{K}$-algebra with identity.

DEFINITION 3.25. *Let $\eta : \mathcal{T} \to \mathcal{T}$ be the antisymmetrizer defined for $u = (X_1, \ldots, X_p)$ by*

$$\eta u = \sum \operatorname{sign} \pi (\pi u) = \sum \operatorname{sign} \pi^{-1} (\pi^{-1} u)$$

summed over all $\pi \in \mathrm{Sym}(p)$.

Define a $\mathbf{K}$-linear map $\lambda : \mathcal{T} \to \mathcal{T}$ by $\lambda 1 = 1$ and

$$\lambda(X_1, \ldots, X_p) = (X_1, X_1 \cap X_2, \ldots, X_1 \cap X_2 \cap \cdots \cap X_p).$$

LEMMA 3.26. *We have*

(i) $\eta(X_1, \ldots, X_p) = (X_1) * \cdots * (X_p)$,

(ii) *if $u, v \in \mathcal{T}$ then $\lambda(\lambda u * \lambda v) = \lambda(u * v)$.*

Proof. Assertion (i) follows by induction. It suffices to check (ii) for $u = (X_1, \ldots, X_p)$ and $v = (Y_1, \ldots, Y_q)$. Then $\lambda u = (X'_1, \ldots, X'_p)$ and $\lambda v = (Y'_1, \ldots, Y'_q)$ where $X'_i = X_1 \cap \cdots \cap X_i$ and $Y'_j = Y_1 \cap \cdots \cap Y_j$. Write $(Z_1, \ldots, Z_{p+q}) = (X_1, \ldots, X_p, Y_1, \ldots, Y_q)$ and $(Z'_1, \ldots, Z'_{p+q}) = (X'_1, \ldots, X'_p, Y'_1, \ldots, Y'_q)$. It follows from the idempotence $Z \cap Z = Z$ that $Z'_{\pi 1} \cap \cdots \cap Z'_{\pi i} = Z_{\pi 1} \cap \cdots \cap Z_{\pi i}$ for all $1 \leq i \leq p + q$ and all (p, q)-shuffles π of $1, \ldots, p + q$. Thus

$$\begin{aligned} \lambda(\lambda u * \lambda v) &= \sum (\operatorname{sign} \pi) \lambda(Z'_{\pi 1}, \ldots, Z'_{\pi(p+q)}) \\ &= \sum (\operatorname{sign} \pi) \lambda(Z_{\pi 1}, \ldots, Z_{\pi(p+q)}) \\ &= \lambda(u * v). \end{aligned}$$

DEFINITION 3.27. *Let $\mathcal{U} = \lambda(\mathcal{T})$. Then $\mathcal{U}$ inherits a grading from $\mathcal{T}$. Since λ is idempotent, $\mathcal{U}$ is spanned by the identity and all $(X_1, \dots, X_p)$ with $X_1 \leq \dots \leq X_p$. Define a product in $\mathcal{U}$ by $uv = \lambda(u * v)$ for $u, v \in \mathcal{U}$.*

The multiplication in $\mathcal{U}$ is associative. To see this, let $u, v, w \in \mathcal{U}$. Since $\lambda w = w$ it follows from Lemma 3.26.ii that

$$(uv)w = \lambda(uv * w) = \lambda(\lambda(u * v) * \lambda w) = \lambda((u * v) * w).$$

The conclusion follows since $*$ is associative. Thus $\mathcal{U}$ is an associative, anticommutative algebra with identity.

Recall the notation $S = (H_1, \dots, H_p)$ and the set $\mathbf{S}$. We may view each element $S \in \mathbf{S}$ as an element of $\mathcal{T}$.

DEFINITION 3.28. *For $S \in \mathbf{S}$ define an element $b_S \in \mathcal{U}$ as follows: if $S = (\,)$ let $b_S = 1$ and for $S = (H_1, \dots, H_p)$ let*

$$b_S = \lambda(\eta S) = \sum_{\pi \in \mathrm{Sym}(p)} \operatorname{sign} \pi (H_{\pi 1}, H_{\pi 1} \cap H_{\pi 2}, \dots, H_{\pi 1} \cap H_{\pi 2} \cap \dots \cap H_{\pi p}).$$

LEMMA 3.29. *Let $S, T \in \mathbf{S}$. Then $b_S b_T = b_{(S,T)}$.*

Proof. Let $S = (H_1, \dots, H_p)$ and $T = (K_1, \dots, K_q)$ where $H_i, K_j \in \mathcal{A}$. Using Lemma 3.26 we get:

$$\begin{aligned} b_S b_T &= \lambda(b_S * b_T) \\ &= \lambda(\lambda(\eta S) * \lambda(\eta T)) \\ &= \lambda(\eta S * \eta T) \\ &= \lambda((H_1) * \dots * (H_p) * (K_1) * \dots * (K_q)) \\ &= \lambda\eta(H_1, \dots, H_p, K_1, \dots, K_q) \\ &= b_{(S,T)}. \end{aligned}$$

DEFINITION 3.30. *Let*

$$B = B(\mathcal{A}) = \sum_{S \in \mathbf{S}} \mathbf{K} b_S.$$

It follows from Lemma 3.29 that $B = \bigoplus_{p \geq 0} B_p$ is a graded subalgebra of $\mathcal{U}$.

LEMMA 3.31. *If $S \in \mathbf{S}$ is dependent then $b_S = 0$. In particular, $b_S = 0$ if $|S| > \ell$ so*

$$B = \bigoplus_{p=0}^{\ell} B_p.$$

Proof. Let $S = (H_1, \dots, H_p)$. If $S_k = (H_1, \dots, \hat{H}_k, \dots, H_p)$ is dependent for some k then $b_S = (-1)^{k-1} b_{(H_k, S_k)} = (-1)^{k-1} b_{H_k} b_{S_k}$ and we are done by induction. Thus we may assume that S_k is independent for each k. It follows that $\bigcap S_k = \bigcap S$ for all k. If $\pi \in \mathrm{Sym}(p)$ let ζ be the permutation defined by $\zeta k = \pi k$ for $1 \le k \le p-2$, $\zeta(p-1) = \pi(p)$ and $\zeta(p) = \pi(p-1)$. Then $\mathrm{sign}\,\zeta = -\mathrm{sign}\,\pi$ and the terms corresponding to π and ζ in Definition 3.28 cancel.

EXAMPLE 3.32. *Suppose $\ell = 2$ and $\mathcal{A} = \{H_1, \dots, H_n\}$. Write $b_k = b_{H_k} = (H_k)$. Then we have*

$$B(\mathcal{A}) = \mathbf{K} \oplus \bigoplus_{p=1}^{n} \mathbf{K} b_p \oplus \bigoplus_{k=1}^{n-1} \mathbf{K} b_k b_n.$$

We know B_0, B_1 and that $B_p = 0$ for $p > 2$. By definition

$$b_i b_j = b_{(H_i, H_j)} = (H_i, H_i \cap H_j) - (H_j, H_j \cap H_i).$$

Thus it is clear that B_2 is spanned by $b_k b_n$ for $1 \le k < n$. It is equally clear that these generators are linearly independent and hence the sum is direct. The reader should compare this example with Example 3.9.

Our aim is to show that $B(\mathcal{A})$ is isomorphic to $A(\mathcal{A})$. Recall the algebra E and note that for $S = (H_1, \dots, H_p)$ and $\pi \in \mathrm{Sym}(p)$ we have $\eta \pi S = (\mathrm{sign}\,\pi)\eta S$ and hence $b_{\pi S} = (\mathrm{sign}\,\pi) b_S$. This allows us to define the following map.

DEFINITION 3.33. *Define a* $\mathbf{K}$*-linear map $\psi : E \to B$ by $\psi e_S = b_S$. Since $e_S e_T = e_{(S,T)}$ the map ψ is a homomorphism of algebras.*

DEFINITION 3.34. *Define a* $\mathbf{K}$*-linear map $\tau : \mathcal{T} \to \mathcal{T}$ by $\tau 1 = 0$, $\tau(X) = 1$ for $X \in L - \{V\}$, and $\tau(X_1, \dots, X_p) = (-1)^{p-1}(X_1, \dots, X_{p-1})$ for $p \ge 2$ and $X_i \in L - \{V\}$.*

LEMMA 3.35. *The following diagram commutes:*

$$\begin{array}{ccc} E_p & \xrightarrow{\partial} & E_{p-1} \\ \psi \downarrow & & \downarrow \psi \\ B_p & \xrightarrow{\tau} & B_{p-1} \end{array}$$

In particular $\tau B_p \subset B_{p-1}$.

Proof. If $S \in \mathbf{S}_p$ then

$$\begin{aligned} \psi \partial e_S &= \lambda\left(\sum_{k=1}^{p} (-1)^{k-1} \eta S_k\right) \\ &= \lambda\left(\sum_{k=1}^{p} \sum_{\zeta \in W_k} (-1)^{k-1} (\mathrm{sign}\,\zeta)(H_{\zeta 1}, \dots, \hat{H}_{\zeta k}, \dots, H_{\zeta p})\right) \end{aligned}$$

where W_k is the group of permutations of $1, \ldots, \hat{k}, \ldots, p$. On the other hand since $\tau\lambda = \lambda\tau$ we have

$$\tau\psi e_S = \lambda\left(\sum \operatorname{sign} \pi (H_{\pi 1}, \ldots, H_{\pi(p-1)})\right)$$

where π ranges over $\operatorname{Sym}(p)$. If $\pi \in \operatorname{Sym}(p)$ and $\pi(p) = k$ define $\zeta \in W_k$ by $\zeta i = \pi i$ for $1 \leq i \leq k-1$, $\zeta i = \pi(i-1)$ for $i > k$. Then $\operatorname{sign} \pi = (-1)^{p-k} \operatorname{sign} \zeta$ and the sums $\psi \partial e_S$ and $\tau\psi e_S$ are equal term for term.

LEMMA 3.36. *The map $\psi : E \to B$ induces a surjection of algebras $\theta : A \to B$ such that $\theta a_S = b_S$.*

Proof. If S is dependent then $\psi \partial e_S = \tau \psi e_S = \tau b_S = 0$ so $\partial e_S \in \ker \psi$. Thus $I \subset \ker \psi$ and ψ induces a surjective map $\theta : A \to B$ such that $\theta a_S = b_S$. Since ψ is an algebra homomorphism, so is θ.

LEMMA 3.37. *The map $\tau : B \to B$ satisfies*
(i) $\tau^2 = 0$,
(ii) *If $b \in B_p$ and $u \in B$ then*

$$\tau(bu) = \tau(b)u + (-1)^p b\tau(u).$$

(iii) *If $\mathcal{A}$ is not empty then the complex (B, τ) is acyclic and hence*

$$\sum_{p=0}^{\ell} (-1)^p \dim B_p = 0.$$

Proof. Properties (i) and (ii) follow from the corresponding facts for ∂_E. Since (B, τ) is acyclic, its Euler characteristic is zero. This implies (iii).

LEMMA 3.38. *If $X \in L$ let $B_X = \sum_{S \in \mathbf{S}_X} \mathbf{K} b_S$. Then*
(i) $B(\mathcal{A}) = \bigoplus_{X \in L(\mathcal{A})} B_X(\mathcal{A})$.
(ii) $B_X(\mathcal{A}_X) \simeq B_X(\mathcal{A})$.

Proof. Assertion (i) is immediate from the definition of B. To prove (ii) note that there is a natural inclusion $\mathcal{T}(\mathcal{A}_X) \to \mathcal{T}(\mathcal{A})$, and because intersections in L_X are the same as in L there is a natural inclusion $\mathcal{U}(\mathcal{A}_X) \to \mathcal{U}(\mathcal{A})$ and hence $B(\mathcal{A}_X) \to B(\mathcal{A})$.

PROPOSITION 3.39. *If $X \in L(\mathcal{A})$ then* $\dim B_X = (-1)^{r(X)} \mu(X)$.

Proof. The argument is the same as for A_X in Proposition 3.20.

THEOREM 3.40. *Let $\mathcal{A}$ be an arrangement. Then $\theta : A(\mathcal{A}) \to B(\mathcal{A})$ is an isomorphism of algebras.*

Proof. Since $\dim A_X = \dim B_X$ for all $X \in L$, the surjective map $\theta : A_X \to B_X$ is an isomorphism for all $X \in L$. Thus $\theta : A \to B$ is an isomorphism.

COROLLARY 3.41. *We have* $B_p = \bigoplus_{X \in L_p} B_X$ *and the Poincaré polynomial of* $B(\mathcal{A})$ *is*

$$P(B(\mathcal{A}), t) = \pi(\mathcal{A}, t).$$

DEFINITION 3.42. *Call the arrangements* $\mathcal{A}_1$ *and* $\mathcal{A}_2$ **algebra equivalent** *if* $A(\mathcal{A}_1)$ *and* $A(\mathcal{A}_2)$ *are isomorphic as graded algebras.*

The question whether the arrangements $\mathcal{A}_1$ and $\mathcal{A}_2$ of Example 2.33 are algebra equivalent was open for several years. Recently Falk [44] used his work on minimal models to find an invariant of the algebra A which is different for these two arrangements. After the conference Rose and Terao [118] constructed two algebra equivalent arrangements which are not lattice isomorphic.

4 Lattice Homology

In this section we associate to the lattice $L(\mathcal{A})$ a simplicial complex $F(\mathcal{A})$, first studied by Folkman [49] and Rota [119], compute its homology groups, and determine its homotopy type. We also associate to $L(\mathcal{A})$ another chain complex, studied by Deheuvels [32] and Baclawski [6], compute its homology and relate it to the homology of $F(\mathcal{A})$ and to the algebra $B(\mathcal{A})$.

There is an active area of research concerned with the topological properties of complexes obtained from partially ordered sets, such as the poset of all subgroups of a group. Since $L(\mathcal{A})$ is a geometric lattice we need not be concerned with the deeper aspects of that theory. We shall present only as much as we need for arrangements. We use books by Dold [35] and Spanier [133] as general references in topology and papers by Rota [119], Folkman [49], Björner [13] and Quillen [113] as general references for combinatorial topology.

The Order Complex

DEFINITION 4.1. *Let P be a partially ordered set. Let $K = K(P)$ be the simplicial complex associated to P as follows:*

(i) *the vertices of K are the elements of P,*

(ii) *a set of vertices $\{X_0, \dots, X_q\}$ spans a q-simplex if and only if it is a linearly ordered subset of P; so after relabeling*

$$X_0 < \cdots < X_q.$$

DEFINITION 4.2. *Given a poset P and the associated simplicial complex $K(P)$, let $|K(P)|$ be the corresponding geometric complex called the* **order complex**.

If P_1 and P_2 are posets then there is a natural partial order on the set $P_1 \times P_2$ given by:

$$(X_1, X_2) \le (Y_1, Y_2) \Leftrightarrow X_1 \le Y_1 \quad \textit{and} \quad X_2 \le Y_2.$$

PROPOSITION 4.3. *Let P be any poset and let $Q = \{0 < 1\}$ be a poset with two elements. Then $|K(P \times Q)|$ is a subdivision of $|K(P)| \times I$.*

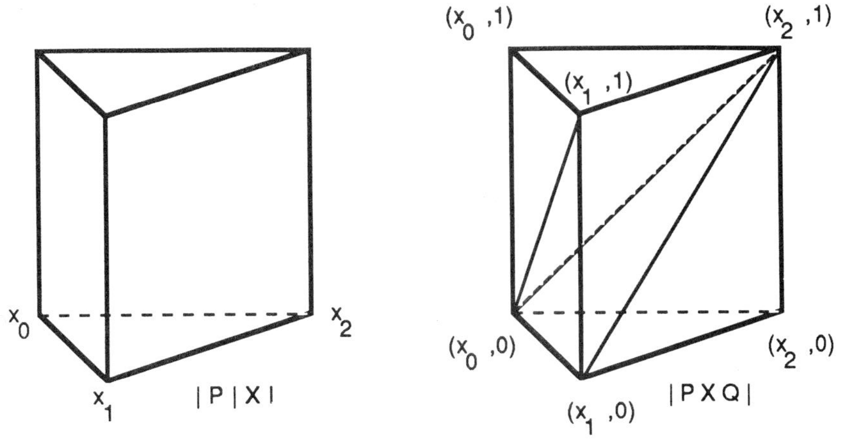

Figure 12: Subdivision of $|\Delta^2| \times I$

Proof. The space $|K(Q)| = I$ is the unit interval. It is sufficient to prove the special case when $P = \{X_0 < \cdots < X_p\}$ is a linearly ordered set, so $|K(P)| = \Delta^p$ is a simplex. Consider Δ^p as a simplex of the standard simplicial subdivision of the p-cube I^p, see [35, p. 118]. The complex $|K(P \times Q)|$ is a simplicial subdivision of the subspace $\Delta^p \times I$ in $I^p \times I = I^{p+1}$. Figure 12 illustrates the case $p = 2$.

COROLLARY 4.4. *Let P be a poset and let $f : P \to P$ be an order preserving map with the property that $f(X) \leq X$ for all $X \in P$. Then the induced map of topological spaces $|f| : |K(P)| \to |K(P)|$ is homotopic to the identity. If $P_0 \subset P$ is a subset such that $f|_{P_0} = \mathrm{id}_{P_0}$ then the homotopy is relative $|K(P_0)|$.*

Proof. Let $Q = \{0 < 1\}$. Define $F : P \times Q \to P$ by $F(X, 0) = f(X)$ and $F(X, 1) = X$. Since $f(X) \leq X$, F is order preserving. It induces a map

$$|F| : |K(P \times Q)| \to |K(P)|.$$

By Proposition 4.3 we may view $|F|$ as a homotopy between $|f|$ and the identity:

$$|F| : |K(P)| \times I \to |K(P)|.$$

If $f|_{P_0} = \mathrm{id}_{P_0}$ then $|F|$ is the identity on $|K(P_0)|$.

LEMMA 4.5. (i) *Suppose P contains a minimal element V. Then $|K(P)|$ is a cone with base $|K(P - \{V\})|$ so $|K(P)|$ is a contractible space.*

(ii) *Suppose P contains a maximal element T. Then $|K(P)|$ is a cone with base $|K(P - \{T\})|$ so $|K(P)|$ is a contractible space.*

Proof. We prove (i), the argument is similar for (ii). If

$$\sigma^p = [X_0, \ldots, X_p] \in K(P - \{V\})$$

then

$$\tau^{p+1} = [V, X_0, \dots, X_p] \in K(P).$$

Moreover, every simplex of $K(P) - K(P - \{V\})$ has the form $V\sigma$ for some $\sigma \in K(P - \{V\})$.

The Folkman Complex

DEFINITION 4.6. *Let $\mathcal{A}$ be an arrangement and let $L = L(\mathcal{A})$. Suppose $r(\mathcal{A}) \geq 2$. Let $K(\mathcal{A}) = K(L - \{V, T\})$ be the simplicial complex associated to the poset obtained from L by deleting its minimal and its maximal element. Let the* **Folkman complex** *$F(\mathcal{A}) = |K(\mathcal{A})|$ be the corresponding geometric complex.*

Note that $\dim F(\mathcal{A}) = r(\mathcal{A}) - 2$. If $r(\mathcal{A}) = 2$ then $F(\mathcal{A})$ consists of $|\mathcal{A}|$ points.

EXAMPLE 4.7. *Let $\mathcal{B}(\ell + 1)$ denote the Boolean arrangement defined by $Q = x_0 x_1 \cdots x_\ell$. Let $H_i = \ker(x_i)$. Then F is the $(\ell - 1)$-complex consisting of the barycentric subdivision of the boundary of an ℓ-simplex with vertices $H_0, \dots, H_\ell$. Thus F is homeomorphic to $S^{\ell-1}$.*

DEFINITION 4.8. *Let $(\mathcal{A}, \mathcal{A}', \mathcal{A}'')$ be a triple with respect to $H_0 \in \mathcal{A}$. Let $L' = L(\mathcal{A}')$, $L'' = L(\mathcal{A}'')$ and $T = T(\mathcal{A})$. Define*

$$F'' = F''(\mathcal{A}) = F(\mathcal{A}'')$$

and

$$F' = F'(\mathcal{A}) = \begin{cases} |K(L' - \{V, T\})| & \text{if } T \in L', \\ |K(L' - \{V\})| & \text{if } T \notin L'. \end{cases}$$

This case distinction is essential in several proofs. Recall from Definition 2.30 that H_0 is a separator if $T \notin L(\mathcal{A}')$. The poset $L' - \{V\}$ has a unique maximal element $T' = T(\mathcal{A}')$. Thus by Lemma 4.5 the space $F'(\mathcal{A})$ is contractible if H_0 is a separator. If H_0 is not a separator then $F'(\mathcal{A}) = F(\mathcal{A}')$.

EXAMPLE 4.9. *Let $\mathcal{A}$ be the 3-arrangement defined by*

$$Q(\mathcal{A}) = xyz(x + y)(x + y + z).$$

Let $H_0 = \ker(x + y + z)$, $H_1 = \ker(x)$, $H_2 = \ker(y)$, $H_3 = \ker(z)$, $H_4 = \ker(x + y)$. The 1-complex $F(\mathcal{A})$ is illustrated in Figure 13. The 0-complex $F''(\mathcal{A})$ consists of the three points $H_0 \cap H_1$, $H_0 \cap H_2$, $H_0 \cap H_3 \cap H_4$. Here $T \in L'$ so H_0 is not a separator. The 1-complex F' is also illustrated in Figure 13.

EXAMPLE 4.10. *In $\mathcal{B}(\ell + 1)$ the complex F'' is homeomorphic to $S^{\ell-2}$. Note here that $T \notin L'$ so H_0 is a separator. The complex F' is the $(\ell - 1)$-simplex opposite the vertex H_0. These complexes are illustrated for $\ell = 3$ in* Figure 14.

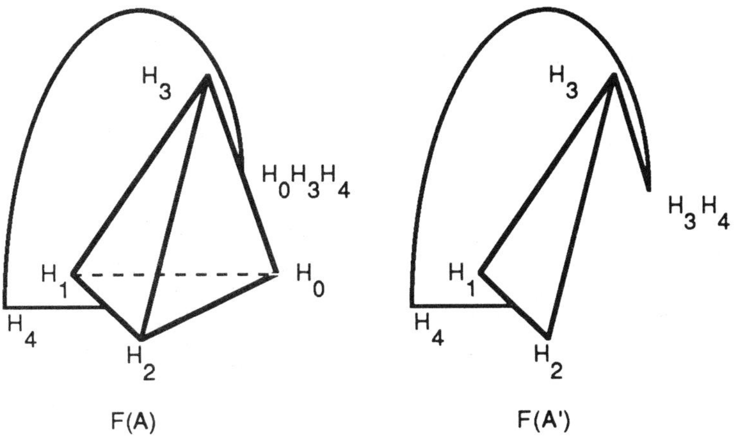

Figure 13: Folkman complexes for $Q = xyz(x+y)(x+y+z)$

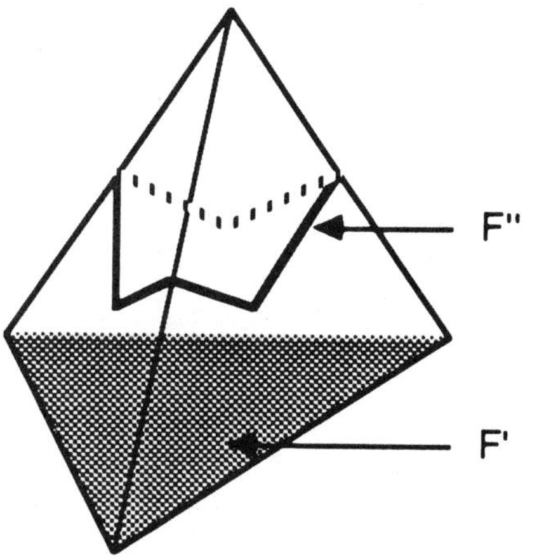

Figure 14: Complexes for the Boolean arrangement

Lemma 4.11. *If $\mathcal{A}$ is an arrangement of rank 2 then $F(\mathcal{A})$ consists of $\mu(\mathcal{A})+1$ points.*

Proof. We observed that F consists of $|\mathcal{A}|$ points and the Möbius function gives $1-|\mathcal{A}|+\mu(\mathcal{A})=0$.

Lemma 4.12. *If $\mathcal{A}$ is an arrangement with $r(\mathcal{A}) \geq 3$ then $F(\mathcal{A})$ is path connected.*

Proof. Suppose X is a vertex of $F(\mathcal{A})$. There exists $H \in \mathcal{A}$ such that $X \geq H$ and $H \in F(\mathcal{A})$. If $X \neq H$ then the 1-simplex $[H,X] \subset F(\mathcal{A})$. Thus every vertex and hence every point of $F(\mathcal{A})$ is connected by a path to some vertex of $F(\mathcal{A})$ which corresponds to a hyperplane. It remains to show that vertices corresponding to distinct hyperplanes $H_1, H_2 \in \mathcal{A}$ are connected. Since $r(H_1 \cap H_2) = 2 < r(\mathcal{A})$ we have $H_1 \cap H_2 \in F(\mathcal{A})$. Thus the 1-simplexes $[H_1, H_1 \cap H_2]$ and $[H_2, H_1 \cap H_2]$ are in $F(\mathcal{A})$.

Let K be a simplicial complex and let $|K|$ be its geometric complex. Let v be a vertex of K. Recall that the **star** of v is a subset of $|K|$ consisting of all open simplexes whose closure contains v:

$$\mathrm{st}(v) = \{|\, \mathring{\sigma}\, | \mid v \in \sigma\}.$$

Note that its closure, $\overline{\mathrm{st}(v)}$ is a cone with cone point v.

Proposition 4.13. *Let $(\mathcal{A}, \mathcal{A}', \mathcal{A}'')$ be a triple with respect to H_0 and let $F = F(\mathcal{A}), F' = F'(\mathcal{A})$. There is a strong deformation retraction*

$$\rho : F - \mathrm{st}(H_0) \to F'.$$

Proof. Note first that $F - \mathrm{st}(H_0) = |K(L - \{V, H_0, T\})|$. Define a poset map

$$\rho : L - \{V, H_0, T\} \to L - \{V, H_0, T\}$$

by

$$\rho(X) = \bigcap_{H \in \mathcal{A}_X - \{H_0\}} H.$$

Note that if H_0 is a separator then $\mathrm{im}(\rho) \subset L' - \{V\}$ and if H_0 is not a separator then $\mathrm{im}(\rho) \subset L' - \{V, T\}$. Extend ρ linearly to $F - \mathrm{st}(H_0)$ and call the resulting map again ρ. It follows that $\mathrm{im}(\rho) \subset F'$. Clearly $r(\rho(X)) \leq r(X)$ and $\rho|_{F'} = \mathrm{id}_{F'}$. Thus by Corollary 4.4 ρ is a strong deformation retraction.

Theorem 4.14. *If $\mathcal{A}$ is an arrangement with $r(\mathcal{A}) \geq 4$ then $F(\mathcal{A})$ is simply connected.*

Proof. We use induction on $|\mathcal{A}|$. Since $|\mathcal{A}| \geq r(\mathcal{A})$, the induction starts with $|\mathcal{A}| = r(\mathcal{A})$. In this case $\mathcal{A}$ is isomorphic to the Boolean arrangement $\mathcal{B}(q)$ for $q = r(\mathcal{A})$. Example 4.7 showed that $F(\mathcal{B}(q))$ is homeomorphic to S^{q-2}. Since $q = r(\mathcal{A}) \geq 4$, the assertion holds for $|\mathcal{A}| = r(\mathcal{A})$. For the induction step choose $H_0 \in \mathcal{A}$ and consider the associated spaces F, F', F''. We have

(1) $$F = \overline{\mathrm{st}(H_0)} \cup (F - \mathrm{st}(H_0))$$

and

(2) $$F'' = \overline{\mathrm{st}(H_0)} \cap (F - \mathrm{st}(H_0)).$$

Now $\overline{\mathrm{st}(H_0)}$ is a cone over F'' with cone point H_0. In particular it is simply connected. We showed in Proposition 4.13 that $F - \mathrm{st}(H_0)$ has the homotopy type of F'. If H_0 is a separator then F' is contractible. If H_0 is not a separator then $F' = F(\mathcal{A}')$ and $r(\mathcal{A}') = r(\mathcal{A})$. Since $|\mathcal{A}'| < |\mathcal{A}|$ the induction hypothesis implies that F' is simply connected. Finally, $r(\mathcal{A}'') = r(\mathcal{A}) - 1 \geq 3$ so it follows from Lemma 4.12 that F'' is path connected. The van Kampen theorem implies that F is simply connected.

Next we want to compute the homology groups and the homotopy type of $F(\mathcal{A})$. Here homology has integer coefficients and $\tilde{H}$ denotes reduced homology. Consider the Mayer-Vietoris sequence of reduced homology for the excisive couple $\{F - \mathrm{st}(H_0), \overline{\mathrm{st}(H_0)}\}$. Using (1) and (2) we get the long exact sequence:

$$\cdots \to \tilde{H}_p(F - \mathrm{st}(H_0)) \oplus \tilde{H}_p(\overline{\mathrm{st}(H_0)}) \overset{(i_1, i_2)_*}{\to} \tilde{H}_p(F)$$
$$\overset{\partial_*}{\to} \tilde{H}_{p-1}(F'') \overset{(j_1, -j_2)_*}{\to} \tilde{H}_{p-1}(F - \mathrm{st}(H_0)) \oplus \tilde{H}_{p-1}(\overline{\mathrm{st}(H_0)}) \to \cdots.$$

The fact that $\overline{\mathrm{st}(H_0)}$ is contractible and Proposition 4.13 give

(3) $$\cdots \to \tilde{H}_p(F') \overset{i_{1*}}{\to} \tilde{H}_p(F) \overset{\partial_*}{\to} \tilde{H}_{p-1}(F'') \overset{j_{1*}}{\to} \tilde{H}_{p-1}(F') \to \cdots.$$

The next result is due to Folkman [49].

THEOREM 4.15. *Let $\mathcal{A}$ be an arrangement with Folkman complex $F = F(\mathcal{A})$. Then*

$$\tilde{H}_i(F) = \begin{cases} 0 & \text{if } i \neq r(\mathcal{A}) - 2, \\ \text{free of rank } |\mu(\mathcal{A})| & \text{if } i = r(\mathcal{A}) - 2. \end{cases}$$

Proof. We use induction on $r(\mathcal{A})$, and for fixed $r(\mathcal{A})$ on $|\mathcal{A}|$. The assertion is correct for $r(\mathcal{A}) = 2$ and arbitrary $|\mathcal{A}|$ by Lemma 4.11. The assertion is also correct for arbitrary $r(\mathcal{A})$ when $|\mathcal{A}| = r(\mathcal{A})$, since in that case $\mathcal{A}$ is the Boolean arrangement and we noted in Example 4.7 that $F(\mathcal{A})$ is an $(r(\mathcal{A}) - 2)$-sphere, while it follows from Proposition 2.20 that $|\mu(\mathcal{A})| = 1$. For the induction step we assume that the result holds for all arrangements $\mathcal{B}$ with $r(\mathcal{B}) < r(\mathcal{A})$ and for all arrangements $\mathcal{B}$ with $r(\mathcal{A}) = r(\mathcal{B})$ and $|\mathcal{B}| < |\mathcal{A}|$. Fix $H_0 \in \mathcal{A}$. Consider the exact sequence (3). For $p \neq r(\mathcal{A}) - 2$ the induction hypothesis implies that $\tilde{H}_p(F') = \tilde{H}_{p-1}(F'') = 0$ and hence $\tilde{H}_p(F) = 0$. For $p = r(\mathcal{A}) - 2$ the induction hypothesis implies that $\tilde{H}_{p-1}(F'')$ is free of rank $|\mu(\mathcal{A}'')|$. If H_0 is not a separator then $F' = F(\mathcal{A}')$ so the induction hypothesis implies that $\tilde{H}_p(F')$ is free of rank $|\mu(\mathcal{A}')|$. Thus $\tilde{H}_p(F)$ is free of rank $|\mu(\mathcal{A}')| + |\mu(\mathcal{A}'')|$. If H_0 is a separator then F' is contractible so $\tilde{H}_p(F)$ is free of rank $|\mu(\mathcal{A}'')|$. The conclusion follows from Corollary 2.31.

DEFINITION 4.16. *Let $(S_1^k, P_1), \ldots, (S_m^k, P_m)$ be m disjoint k-spheres with base points P_j. Their* **wedge** *is the based space $(\bigvee_m S^k, P)$ obtained by identifying the base points $P_1 = \cdots = P_m = P$.*

It is clear that $(\bigvee_m S^k, P)$ is a CW-complex. We may write $\bigvee_m S^k$ for brevity. We have

$$\pi_i(\bigvee_m S^k) = \begin{cases} 0 & \text{for } i < k, \\ \text{free of rank } m & \text{for } i = k, \end{cases}$$

and

$$\tilde{H}_i(\bigvee_m S^k; \mathbf{Z}) = \begin{cases} 0 & \text{for } i \neq k, \\ \text{free of rank } m & \text{for } i = k. \end{cases}$$

The next result was stated by Quillen [113] without proof. It is safe to assume that he had in mind the argument below. Björner and Walker [16] proved the result without appeal to facts in homotopy theory.

THEOREM 4.17. *Let $\mathcal{A}$ be an arrangement with $r(\mathcal{A}) \geq 2$. Then its Folkman complex $F = F(\mathcal{A})$ has the homotopy type of $\bigvee_m S^k$ with $k = r(\mathcal{A}) - 2$ and $m = |\mu(\mathcal{A})|$.*

Proof. For $r(\mathcal{A}) = 2$ this follows from Lemma 4.11. For $r(\mathcal{A}) = 3$ the complex F is 1-dimensional and hence it has the homotopy type of a wedge of circles. Their number equals the rank of $H_1(F)$. We showed in Theorem 4.15 that this rank is m. For $r(\mathcal{A}) \geq 4$ the complex F is simply connected by Theorem 4.14. It follows from Theorem 4.15 and the Hurewicz isomorphism theorem [133, p. 398] that $\pi_i(F) = 0$ for $1 \leq i < k$ and $\pi_k(F) \approx H_k(F; \mathbf{Z})$. The last group is free of rank m by Theorem 4.15. For $1 \leq i \leq m$ let $\rho_i : S_i^k \to F$ be generators of $\pi_k(F)$. Let $\rho : \bigvee_m S^k \to F$ be the sum of ρ_i. Then ρ induces an isomorphism in homology by construction. Since the spaces are simply connected CW-complexes, it follows from standard results in homotopy theory [133, pp. 405-406] that ρ is a homotopy equivalence.

Whitney Homology

We close this section with a construction due to Deheuvels [32] and Baclawski [6]. Recall the spaces $\mathcal{T}_p$ from Section 3.

DEFINITION 4.18. *Let $\mathcal{A}$ be an arrangement with lattice $L = L(\mathcal{A})$. Define a chain complex $(\mathcal{C}, \delta)$ as follows. Let $\mathcal{C}_0 = \mathbf{K}$ and for $p > 0$ let $\mathcal{C}_p \subset \mathcal{T}_p$ have a $\mathbf{K}$-basis consisting of all p-tuples $(X_1, \dots, X_p)$ where $X_i \in L - \{V\}$ and*

$$X_1 < \cdots < X_p.$$

Let $\mathcal{C} = \bigoplus_{p=0}^{\ell} \mathcal{C}_p$. Define a $\mathbf{K}$-linear map $\delta : \mathcal{C} \to \mathcal{C}$ by $\delta(1) = 0$, $\delta(X) = 0$ for $X \in L - \{V\}$, and for $p \geq 2$

$$\delta(X_1, \dots, X_p) = \sum_{k=1}^{p-1} (-1)^{k-1} (X_1, \dots, \hat{X}_k, \dots, X_p).$$

Note that δ differs from the usual boundary operator in that X_p is never deleted. We still have $\delta^2 = 0$ so $(\mathcal{C}, \delta)$ is a chain complex. We call it the **Whitney complex** of $\mathcal{A}$. The Poincaré polynomial of its homology was first computed by Baclawski [6]. We give an elementary proof of his result.

PROPOSITION 4.19. *Let $\mathcal{A}$ be an arrangement. Let $\mathcal{H} = \mathcal{H}(\mathcal{A})$ be the homology of $(\mathcal{C}, \delta)$. The Poincaré polynomial of $\mathcal{H}(\mathcal{A})$ is*

$$P(\mathcal{H}(\mathcal{A}), t) = \pi(\mathcal{A}, t).$$

Proof. For $X \in L - \{V\}$ let $\mathcal{C}_X$ be the subspace of $\mathcal{C}$ spanned by all $(X_1, \dots, X_p)$ with $X_p = X$ and let $\mathcal{C}_V = \mathbf{K}$. Then $\mathcal{C} = \bigoplus_{X \in L} \mathcal{C}_X$ and $\delta : \mathcal{C}_X \to \mathcal{C}_X$. Thus $(\mathcal{C}_X, \delta)$ is a subcomplex. Let $\mathcal{H}_X$ be its homology. Then $\mathcal{H} = \bigoplus_{X \in L} \mathcal{H}_X$. Let $\mathcal{C}'_0 = \mathbf{K}$. For $p > 0$ let $\mathcal{C}'_p \subset \mathcal{C}_p$ be the subspace spanned by all $(X_1, \dots, X_p)$ with $X_p \neq T(\mathcal{A})$ and let $\mathcal{C}''_p \subset \mathcal{C}_p$ be the subspace spanned by all $(X_1, \dots, X_p)$ with $X_p = T(\mathcal{A})$. Then $\mathcal{C}_p = \mathcal{C}'_p \oplus \mathcal{C}''_p$. The $\mathbf{K}$-linear map defined by $1 \to T(\mathcal{A})$ and $(X_1, \dots, X_p) \to (X_1, \dots, X_p, T(\mathcal{A}))$ establishes an isomorphism $\mathcal{C}'_p \simeq \mathcal{C}''_{p+1}$ of vector spaces for $0 \leq p < \ell$. Thus the Euler characteristic of $\mathcal{C}$ is zero. Since $\mathcal{H} = \bigoplus_{X \in L} \mathcal{H}_X$ and the natural identification $\mathcal{C}_X(\mathcal{A}_X) \simeq \mathcal{C}_X(\mathcal{A})$ implies $\mathcal{H}_X(\mathcal{A}_X) \simeq \mathcal{H}_X(\mathcal{A})$, we may argue as in Proposition 3.20 that $\dim \mathcal{H}_X = (-1)^{r(X)} \mu(X)$. This proves the assertion.

THEOREM 4.20. (i) *The elements of B are cycles of $\mathcal{C}$ so $\delta B = 0$.*

(ii) *The map $B(\mathcal{A}) \to \mathcal{H}(\mathcal{A})$ which sends b_S to its homology class $[b_S]$ is an isomorphism of graded vector spaces.*

Proof. (i) It suffices to show that $\delta b_S = 0$ for all $S \in \mathbf{S}$. Let $S = (H_1, \dots, H_p)$. In $\delta b_{(H_1, \dots, H_p)}$ each term

$$(\operatorname{sign} \pi)(-1)^{k-1} (H_{\pi 1}, \dots, \widehat{H_{\pi 1} \cap \cdots \cap H_{\pi k}}, \dots, H_{\pi 1} \cap \cdots \cap H_{\pi p})$$

is cancelled against the term in which πk and $\pi(k+1)$ are transposed. For (ii) it suffices to show that the map is a monomorphism. Thus it suffices to show that $B_X \to \mathcal{H}_X$ is a monomorphism. By induction it suffices to show this for $X = T(\mathcal{A})$, where it is obvious since $\mathcal{C}_{r(\mathcal{A})+1} = 0$.

The two constructions of this section are closely related. In fact the homology groups of the Whitney complex equal direct sums of the homology groups of all Folkman subcomplexes of $\mathcal{A}$ in the following sense.

THEOREM 4.21. *Let $\mathcal{A}$ be an arrangement. Then*

$$\mathcal{H}_0(\mathcal{A}) = \mathbf{K}, \qquad \mathcal{H}_1(\mathcal{A}) = \bigoplus_{H \in \mathcal{A}} \mathbf{K}(H),$$

and for $p \geq 2$ the map $\tau : \mathcal{T} \to \mathcal{T}$ induces isomorphisms

$$\mathcal{H}_p(\mathcal{A}) \simeq \bigoplus_{X \in L_p(\mathcal{A})} \tilde{H}_{p-2}(F(\mathcal{A}_X)).$$

Proof. Recall that we identified B_p with the group of p-cycles of $\mathcal{C}$. Since $B_p = \bigoplus_{X \in L_p} B_X$ and $B_X(\mathcal{A}) = B_X(\mathcal{A}_X)$ it suffices to prove the assertion for $X = T(\mathcal{A})$ and $r(\mathcal{A}) \geq 2$. Write $T = T(\mathcal{A})$ and $r(\mathcal{A}) = \ell$. If $\ell = 2$ then

$$B_T \simeq \left(\bigoplus_{H \in \mathcal{A}} \mathbf{K}(H) \right) \Big/ \mathbf{K} \left(\sum_{H \in \mathcal{A}} H \right) \simeq \tilde{H}_0(F(\mathcal{A})).$$

Assume $\ell \geq 3$. Since there are no $(\ell-2)$-boundaries we have $\tilde{H}_{\ell-2}(F) = H_{\ell-2}(F) = Z_{\ell-2}(F)$. We identify the cycle group $Z_{\ell-2}$ with a subspace of $\mathcal{T}_{\ell-1}$. Note that $B_T \subset \mathcal{T}_\ell$. We show that $\tau B_T \subset Z_{\ell-2}(F)$. Let $S = (H_1, \ldots, H_\ell) \in \mathbf{S}_\ell$. If S is dependent then $\tau b_S = \tau 0 = 0$. Suppose S is independent. Let $\mathcal{B}$ denote the subarrangement of $\mathcal{A}$ whose elements are the hyperplanes of S. This is a fine but useful distinction: $\mathcal{B}$ is a set, S is an ordered set with the same elements. Then $L(\mathcal{B})$ is a Boolean lattice. We showed that $H_{\ell-2}(F(\mathcal{B}))$ is one dimensional, generated by the cycle

$$z_S = \sum_{\pi \in \mathrm{Sym}(\ell)} (-1)^{\ell-1} (\operatorname{sign} \pi)(H_{\pi 1}, H_{\pi 1} \cap H_{\pi 2}, \ldots, H_{\pi 1} \cap \cdots \cap H_{\pi(\ell-1)}).$$

Since $\tau b_S = z_S$ this shows that $\tau B_T \subset Z_{\ell-2}(F)$. Let $\mathcal{T}'_\ell$ be the subspace of $\mathcal{T}_\ell$ spanned by all $(X_1, \ldots, X_\ell)$ with $X_\ell = T$. If $S \in \mathbf{S}_\ell$ is independent then $\bigcap \pi S = \bigcap S = T$ for all $\pi \in \mathrm{Sym}(\ell)$ so $B_T \subset \mathcal{T}'_\ell$. But $\tau : \mathcal{T}'_\ell \to \mathcal{T}_{\ell-1}$ is a monomorphism and thus $\tau : B_T \to Z_{\ell-2}(F)$ is a monomorphism. It is an isomorphism because the spaces have the same dimension $(-1)^\ell \mu(\mathcal{A})$. This completes the argument.

5 The Complement $M(\mathcal{A})$

In this section we prove some elementary facts about the topology of the complement of an arrangement over the complex numbers. In addition we outline some of the basic results of Arnold [3], Brieskorn [20], Deligne [33] and Hattori [64] which started the recent activity in the area.

PROPOSITION 5.1. *The complement $M = M(\mathcal{A})$ of an ℓ-arrangement $\mathcal{A}$ is an open, smooth, parallelizable manifold of real dimension 2ℓ, which has the homotopy type of a finite CW-complex.*

Proof. The vector space V is an open, smooth, parallelizable manifold of real dimension 2ℓ and M is open in V. The last assertion is clear if $\mathcal{A}$ is empty. If $\mathcal{A}$ is nonempty let $Q = Q(\mathcal{A})$ be a defining polynomial for $\mathcal{A}$. The map $Q : M \to \mathbf{C}^*$ is the projection of a smooth fiber bundle. It follows from work of Milnor [93] that the restriction of Q to a suitable neighborhood of the origin is a fibration. Since Q is homogeneous we may take this neighborhood to be all of M. Milnor [93] proved that the fiber $F = Q^{-1}(1)$ has the homotopy type of a finite CW-complex. It follows that the same holds for M.

REMARK 5.2. *The map $Q : M \to \mathbf{C}^*$ is the Milnor fibration and the fiber $F = Q^{-1}(1)$ is called the* **Milnor fiber**. *There is very little known about the topology of the Milnor fiber of an arrangement.*

In Section 1 we described another fibration with total space M, which explains the connection between central and affine arrangements. Recall the canonical bundle $p : \mathbf{C}^\ell - \{0\} \to \mathbf{CP}^{\ell-1}$ with fiber $\mathbf{C}^*$, which identifies z with λz for $\lambda \in \mathbf{C}^*$.

PROPOSITION 5.3. *Let $\mathcal{A}$ be a nonempty arrangement with complement $M = M(\mathcal{A})$ and let $M^* = p(M)$. The restriction $p : M \to M^*$ is a trivial fibration so $M = M^* \times \mathbf{C}^*$. There is an affine $(\ell - 1)$-arrangement $\mathcal{A}^*$ such that $M^* = M(\mathcal{A}^*)$.*

Proof. Let $H \in \mathcal{A}$. The restriction of p to $M_H = V - \{H\}$ has base space $\mathbf{CP}^{\ell-1} - \mathbf{CP}^{\ell-2} \approx \mathbf{C}^{\ell-1}$. Thus $p : M_H \to \mathbf{C}^{\ell-1}$ is a trivial bundle and

$p: M \to M^*$ is a subbundle. Choose a basis for V^* such that $H = \ker x_\ell$. Let $Q = Q(\mathcal{A})$ and note that x_ℓ divides Q. Define an affine arrangement $\mathcal{A}^*$ by the polynomial Q^* obtained by setting $x_\ell = 1$ in Q. This is the natural identification $M(\mathcal{A}^*) \approx M^*$.

REMARK 5.4. *The Milnor fiber F admits a free action by the cyclic group of order $n = |\mathcal{A}|$. The quotient space is naturally identified with M^*.*

The map $p: M \to M^*$ is the orbit map of the standard $\mathbf{C}^*$-action

$$t(x_1, \ldots, x_\ell) = (tx_1, \ldots, tx_\ell).$$

Let $G(n)$ denote the cyclic subgroup of $\mathbf{C}^*$ of order n. Since Q is homogeneous of degree n, $G(n)F \subset F$. It follows that $M^* = M/\mathbf{C}^* = F/G(n)$. Let $\zeta = e^{2\pi i/n}$ be a generator of $G(n)$. The map $F \to F$ induced by multiplication by ζ is called the monodromy of the Milnor fiber. Let $\pi : F \to F/G(n)$ and let also $\pi : \mathbf{C}^* \to \mathbf{C}^*/G(n)$. The commutative diagram below connects the two fibrations. Here τ denotes the identity map.

$$\begin{array}{ccccc} & & F & \xrightarrow{\pi} & F/G(n) \\ & & \downarrow i & & \downarrow \tau \\ \mathbf{C}^* & \to & M & \xrightarrow{p} & M^* \\ \downarrow \pi & & \downarrow Q & & \\ \mathbf{C}^* & \xrightarrow{\tau} & \mathbf{C}^* & & \end{array}$$

$K(\pi, 1)$-arrangements

EXAMPLE 5.5. *Let $\mathcal{A}$ be the arrangement of* Example 1.6 *defined by $Q(\mathcal{A}) = xy(x+y)$. The complement M has the homotopy type of $(S^1 \vee S^1) \times S^1$.*

We use Proposition 5.3. If we let $H = \ker y$ go to the line at infinity, then the corresponding affine 1-arrangement is defined by $Q^* = x(x+1)$. Thus M^* is the complement of the two points $x = 0$ and $x = -1$ in $\mathbf{C}$. It follows that M^* has the homotopy type of $S^1 \vee S^1$. Since $\mathbf{C}^*$ has the homotopy type of S^1, the conclusion follows from Proposition 5.3. Note that $\pi_i(M) = 0$ for $i \geq 2$ so M is a $K(\pi, 1)$-space.

DEFINITION 5.6. *Call $\mathcal{A}$ a $K(\pi, 1)$-***arrangement** *if $M(\mathcal{A})$ is a $K(\pi, 1)$-space.*

PROPOSITION 5.7. *Every central 2-arrangement is $K(\pi, 1)$.*

Proof. The argument in Example 5.5 shows that if $\mathcal{A}$ is a central 2-arrangement with $|\mathcal{A}| = n$, then $M(\mathcal{A})$ has the homotopy type of $(\bigvee_{n-1} S^1) \times S^1$.

PROPOSITION 5.8. *The Boolean arrangement is $K(\pi, 1)$.*

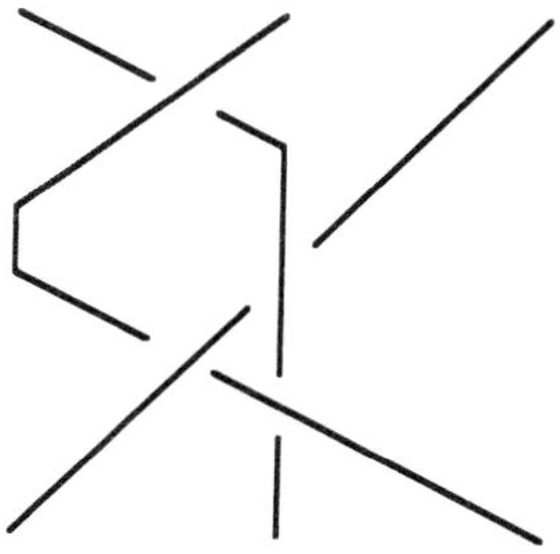

Figure 15: A braid on 3 strands

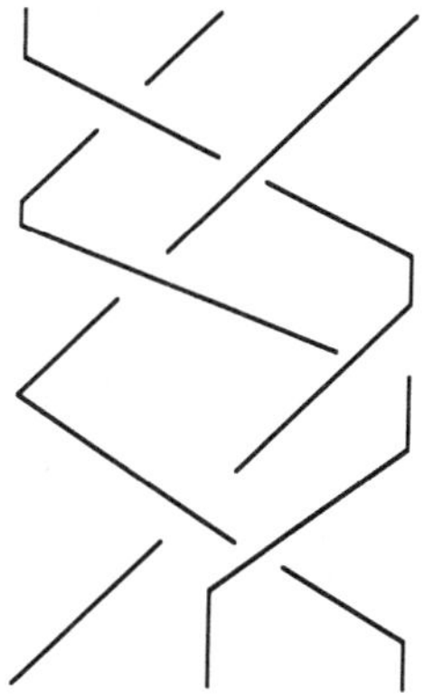

Figure 16: A pure braid on 3 strands

Proof. Since the Boolean arrangement is defined by $Q = x_1 \cdots x_\ell$, M is the complement of the coordinate hyperplanes. Thus $M = (\mathbf{C}^*)^\ell$ which has the homotopy type of an ℓ-torus.

REMARK 5.9. *More on the braid arrangement.*

Before we show that the braid arrangement is also $K(\pi, 1)$, it is appropriate to justify its name and describe some of its history. Braids and the braid group were defined by Artin [5]. Figure 15 shows a braid on 3 strands. Braids with ℓ strands may be composed by juxtaposition. There is a suitable notion of isotopy of braids which makes this an associative multiplication. The inverse of a braid is the braid which untangles it. Isotopy classes of braids on ℓ strands form a group called the **braid group**, $B(\ell)$. For details see Birman's book [12]. There is a natural surjection $B(\ell) \to \mathrm{Sym}(\ell)$ which sends each braid to the permutation of its ends. The image of the braid in Figure 15 is the 3-cycle (1,2,3).

The kernel of this map is called the pure braid group $PB(\ell)$. The corresponding pure braids have the property that each strand returns to its point of origin. Figure 16 shows a pure braid on 3 strands. The braid group is

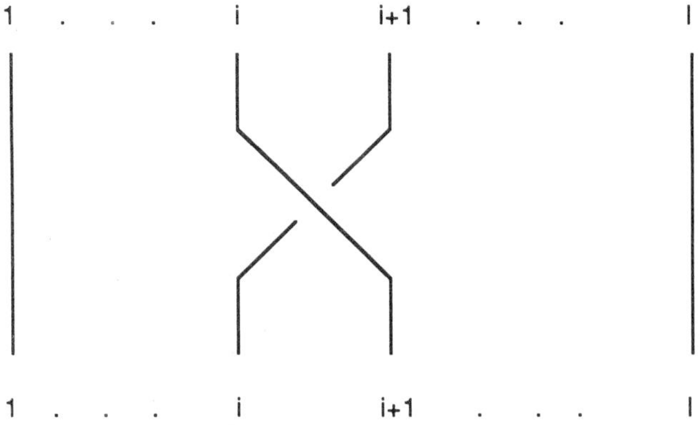

Figure 17: The generator a_i

generated by the braids a_i for $1 \le i < \ell$ of Figure 17 and it is known [12, p.11] that the following relations are sufficient to give a presentation:

$$\begin{aligned} a_i a_j &= a_j a_i && \text{if } |i - j| \ge 2, \\ a_i a_{i+1} a_i &= a_{i+1} a_i a_{i+1} && 1 \le i \le \ell - 2. \end{aligned}$$

The fact that the pure braid group is the fundamental group of the pure braid space as we described it in Section 1 first appeared in a paper by Fox and Neuwirth [50]. To make this statement precise let $\mathcal{A}_\ell$ denote the braid arrangement of Definition 1.10 complexified. Let $M_\ell = M(\mathcal{A}_\ell)$. We described in Section 1 how a pure braid gives rise to a map of the circle into M_ℓ, and hence to an element of its fundamental group. For the converse choose a base point $x \in M_\ell$. An element of $\pi_1(M_\ell, x)$ is represented by a map $f : (I, \{0, 1\}) \to (M_\ell, x)$ which we may assume to be a smooth embedding. The coordinate functions of f are the strands of the pure braid. Thus $\pi_1(M_\ell) = PB(\ell)$. Note that M_ℓ admits a free action of $\mathrm{Sym}(\ell)$ by permuting the coordinates. Let $B_\ell = M_\ell/\mathrm{Sym}(\ell)$ be the orbit space and let $p : M_\ell \to B_\ell$ be the projection of this covering. A similar argument shows that $\pi_1(B_\ell) = B(\ell)$.

The covering $p : M_\ell \to B_\ell$ is a fibration with discrete fiber F of cardinality $|\mathrm{Sym}(\ell)|$. The homotopy long exact sequence of this fibration breaks up into:

$$1 \to \pi_1(M_\ell) \to \pi_1(B_\ell) \to \pi_0(F) \to 1$$

which we may identify with

$$1 \to PB(\ell) \to B(\ell) \to \mathrm{Sym}(\ell) \to 1$$

and for $k \ge 2$ it gives isomorphisms $\pi_k(M_\ell) \approx \pi_k(B_\ell)$. Fadell and Neuwirth [39] showed that M_ℓ is a $K(\pi, 1)$-space.

THEOREM 5.10. *The braid arrangement $\mathcal{A}_\ell$ is $K(\pi, 1)$.*

Proof. The projection map $\mathbf{C}^\ell \to \mathbf{C}^{\ell-1}$ defined by $(x_1, \dots, x_\ell) \to (x_1, \dots, x_{\ell-1})$ induces a locally trivial fibration $M_\ell \to M_{\ell-1}$. The fiber over

$(\xi_1, \ldots, \xi_{\ell-1})$ is $\mathbf{C} - \{\xi_1, \ldots, \xi_{\ell-1}\}$. Thus the fiber retracts onto a wedge of $(\ell - 1)$ circles, so it is a $K(\pi, 1)$-space. Since $M_2 = \{(x_1, x_2) | x_1 \neq x_2\} = \mathbf{C} \times \mathbf{C}^*$ we are done by induction.

The representation of M as the total space of a sequence of fibrations is an imporant tool. The next two definitions and the following proposition are due to Falk and Randell [45]. In [147] Terao proved the existence of a fibration from the existence of modular elements in $L(\mathcal{A})$.

DEFINITION 5.11. *Let $\mathcal{A}$ be an ℓ-arrangement. Call $M(\mathcal{A})$* **strictly linearly fibered** *if after a suitable linear change of coordinates the restriction of the projection to the first $(\ell - 1)$ coordinates is a fiber bundle projection whose base space N is the complement of an arrangement in $\mathbf{C}^{\ell-1}$, and whose fiber is the complex line $\mathbf{C}$ with finitely many points removed.*

DEFINITION 5.12. (i) *The 1-arrangement $\{0\}$ is* **fiber type**.

(ii) *For $\ell \geq 2$ the ℓ-arrangement is* **fiber type** *if $M(\mathcal{A})$ is strictly linearly fibered with base $N = M(\mathcal{B})$ and $\mathcal{B}$ is an $(\ell - 1)$-arrangement of fiber type.*

PROPOSITION 5.13. *A fiber type ℓ-arrangement $\mathcal{A}$ is $K(\pi, 1)$.*

Proof. It follows from the definition that there exist k-arrangements $\mathcal{A}_k$ for $1 \leq k \leq \ell$ with $\mathcal{A} = \mathcal{A}_\ell$ and a sequence of fibrations

$$M(\mathcal{A}_\ell) \xrightarrow{p_\ell} M(\mathcal{A}_{\ell-1}) \xrightarrow{p_{\ell-1}} \cdots \xrightarrow{p_3} M(\mathcal{A}_2) \xrightarrow{p_2} M(\mathcal{A}_1) = \mathbf{C}^*$$

with the fiber F_k of p_k homeomorphic to $\mathbf{C}$ with d_k points removed. The conclusion follows by repeated application of the homotopy exact sequence of a fibration.

The arrangement $\mathcal{A}_1$ of Example 2.33 is strictly linearly fibered. In fact it is easy to see from Figure 10 that $M(\mathcal{A}_1)$ is homeomorphic to $C_3 \times C_3 \times C_1$, where C_k denotes the complex line with k points removed.

In a 1971 Bourbaki Seminar talk [20] Brieskorn generalized Arnold's results. He replaced the symmetric group and the braid arrangement by a Coxeter group W acting in an ℓ-dimensional real vector space $V_{\mathbf{R}}$. Let V be the complexification of $V_{\mathbf{R}}$. Then W acts as a reflection group in V. Let $\mathcal{A} = \mathcal{A}(W)$ be its reflection arrangement. Brieskorn conjectured that $\mathcal{A}(W)$ is a $K(\pi, 1)$ arrangement for all Coxeter groups W. He proved this for some of the groups by representing M as the total space of a sequence of fibrations. Some of these fibrations are not strictly linear. Deligne [33] settled the question by proving the much stronger result stated below.

DEFINITION 5.14. *Let $(\mathcal{A}_{\mathbf{R}}, V_{\mathbf{R}})$ be a real arrangement. Call $\mathcal{A}_{\mathbf{R}}$ a* **simplicial** *arrangement if every component of $M(\mathcal{A}_{\mathbf{R}})$ is an open simplicial cone.*

THEOREM 5.15. *Let* $(\mathcal{A}_{\mathbf{R}}, V_{\mathbf{R}})$ *be a simplicial arrangement. Then its complexification* $(\mathcal{A}, V)$ *is* $K(\pi, 1)$.

This result proves Brieskorn's conjecture because the arrangement of a Coxeter group is simplicial [18]. There exist unitary reflection groups [130] which are not Coxeter groups. It is reasonable to ask if their reflection arrangements are $K(\pi, 1)$. For a subclass of unitary reflection groups called Shephard groups this was proved in [107]. We give an outline of the argument in Section 10. The conjecture is still open for the remaining unitary reflection groups. The arrangement $\mathcal{A}_1$ of Example 2.33 shows that Theorem 5.13 is not a consequence of Theorem 5.15.

Generic Arrangements

So far we have listed a number of ways for an arrangement to be $K(\pi, 1)$. However, this is not generic behavior. In fact it follows from a theorem of Hattori [64] which we state below, that most arrangements are not $K(\pi, 1)$. First we need two definitions.

DEFINITION 5.16. *The arrangement $\mathcal{A}$ is called a* **general position** *arrangement if for every subset* $\{H_1, \ldots, H_p\} \subset \mathcal{A}$ *with* $p \leq \ell$

$$r(H_1 \cap \cdots \cap H_p) = p$$

and when $p > \ell$

$$H_1 \cap \cdots \cap H_p = \emptyset.$$

Note that if $\mathcal{A}$ is a central general position ℓ-arrangement then $|\mathcal{A}| \leq \ell$. Thus the only interesting general position arrangements are affine.

DEFINITION 5.17. *Let* $\mathbf{n} = \{1, \ldots, n\}$. *If* $I \subset \mathbf{n}$ *let* $|I|$ *be its cardinality. Define the subtorus* T_I *of* T^n *by*

$$T_I = \{(z_1, \ldots, z_n) \in T^n | z_j = 1 \text{ for } j \notin I\}.$$

THEOREM 5.18. *Let* $\mathcal{A}^*$ *be an affine ℓ-arrangement in general position and assume that* $n = |\mathcal{A}^*| \geq \ell + 1$. *Then* $M^* = M(\mathcal{A}^*)$ *has the homotopy type of*

$$M_0^* = \bigcup_{|I|=\ell} T_I.$$

Hattori [64] also proved that $\pi_1(M^*)$ is free abelian of rank n, and that the universal covering space $\tilde{M}^*$ of M^* has trivial homology in dimensions $\neq 0, \ell$. He also gave a free $\mathbf{Z}(\pi_1 M^*)$ resolution of $H_\ell(\tilde{M}^*, \mathbf{Z})$. In particular if $n = \ell + 1$ then $H_\ell(\tilde{M}^*, \mathbf{Z})$ is a free $\mathbf{Z}(\pi_1 M^*)$-module of rank 1.

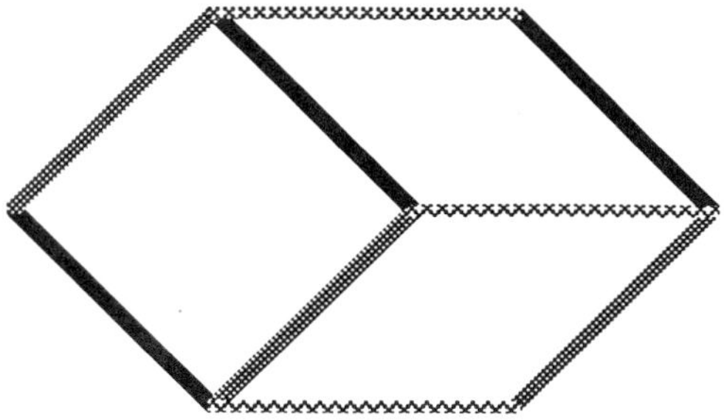

Figure 18: Three lines in general position

DEFINITION 5.19. *Let $\mathcal{A}$ be a central $(\ell + 1)$-arrangement with $\ell \geq 2$. Call $\mathcal{A}$ a* **generic** *arrangement if the hyperplanes of every subarrangement $\mathcal{B} \subseteq \mathcal{A}$ with $|\mathcal{B}| = \ell + 1$ are linearly independent. Equivalently, if $\mathcal{A}^*$ is the corresponding affine ℓ-arrangement then $\mathcal{A}^*$ is a general position arrangement.*

COROLLARY 5.20. *Generic arrangements are not* $\mathbf{K}(\pi, 1)$. $n \geq \ell+1$

Proof. Since $\tilde{M}^*$ is simply connected, it follows from Hattori's results and the Hurewicz isomorphism theorem that $\pi_i(\tilde{M}^*) = H_i(\tilde{M}^*; \mathbf{Z}) = 0$ for $1 \leq i < \ell$ and $\pi_\ell(\tilde{M}^*) = H_\ell(\tilde{M}^*; \mathbf{Z})$. By Hattori's theorem $\pi_\ell(M^*) \neq 0$ and M^* is not a $K(\pi, 1)$-space. The conclusion follows from the fact that $\pi_i(M) = \pi_i(M^*)$ for $i \geq 2$.

EXAMPLE 5.21. *Define the generic arrangement $\mathcal{A}$ by $Q(\mathcal{A}) = xyz(x+y-z)$. Then $\mathcal{A}$ is not a $K(\pi, 1)$ arrangement.*

By Proposition 5.3 we have $M = M^* \times \mathbf{C}^*$, where the affine arrangement $\mathcal{A}^*$ is defined by $Q^* = xy(x+y-1)$. This is the arrangement in Example 1.7. Clearly $\mathcal{A}^*$ is a general position arrangement, so we may use Theorem 5.18. Here $n = 3$ and $\ell = 2$. Thus

$$M_0^* = S^1 \times S^1 \times 1 \cup S^1 \times 1 \times S^1 \cup 1 \times S^1 \times S^1 \subset S^1 \times S^1 \times S^1.$$

We can visualize M_0^* as the identification space obtained from the boundary of a cube by identifying opposite faces and some of their edges. Figure 18 shows the edge identifications on the front three faces of the cube. Each face becomes a torus. Since $n = 3 = \ell + 1$ it follows that $\pi_2(M^*) = \mathbf{Z}$. A nontrivial element of $\pi_2(M_0^*) = \pi_2(M)$ is obtained by any map which sends S^2 onto the boundary of the cube, followed by the identifications.

Arnold's Conjectures

The main result of Arnold's paper [3] was the calculation of the Poincaré polynomial of the pure braid space M_ℓ and the cohomology ring structure of

$H^*(M_\ell)$. Arnold showed that

$$P(M_\ell, t) = (1+t)(1+2t)\cdots(1+(\ell-1)t).$$

The reader should compare this formula with the Poincaré polynomial of the braid arrangement computed in Proposition 2.26. Arnold also showed that $H^*(M_\ell)$ is generated by the 1-dimensional elements

$$\omega_{p,q} = \frac{1}{2\pi i}\frac{dz_p - dz_q}{z_p - z_q}$$

and that all relations among these generators are consequences of the relations:

$$\omega_{p,q}\omega_{q,r} + \omega_{q,r}\omega_{r,p} + \omega_{r,p}\omega_{p,q} = 0.$$

Arnold stated two conjectures for an arbitrary arrangement $\mathcal{A}$. The first said that $H^*(M, \mathbf{Z})$ is torsion free. The second may be stated as follows. Define holomorphic differential forms $\omega_H = (1/2\pi i)(d\alpha_H/\alpha_H)$ for $H \in \mathcal{A}$ and let $[\omega_H]$ denote the corresponding cohomology class. Let $R = \bigoplus_{p=0}^{\ell} R_p$ be the graded $\mathbf{C}$-algebra of holomorphic differential forms on M generated by the ω_H and 1. Arnold conjectured that the natural map $\eta \to [\eta]$ of $R \to H^*(M, \mathbf{C})$ is an isomorphism of graded algebras. This was proved by Brieskorn [20], who showed in fact that the $\mathbf{Z}$-subalgebra of R generated by the forms ω_H and 1 is isomorphic to the singular cohomology $H^*(M, \mathbf{Z})$. In [100] it was shown that for an arbitrary arrangement $\mathcal{A}$ the Poincaré polynomial of $M(\mathcal{A})$ equals the Poincaré polynomial of $\mathcal{A}$. We prove this in Section 6. The structure of the algebra R was also obtained in [100]. This work is presented in Section 7.

Call the arrangements $\mathcal{A}_1 = (\mathcal{A}_1, V)$ and $\mathcal{A}_2 = (\mathcal{A}_2, V)$ **diffeomorphic**, **homeomorphic**, or **homotopy equivalent**, if $M(\mathcal{A}_1)$ and $M(\mathcal{A}_2)$ are diffeomorphic, homeomorphic, or homotopy equivalent. The main unsolved problems ask how these topological equivalence classes relate to the combinatorial equivalence classes defined earlier. From this point of view the result of the next section says that two arrangements have isomorphic cohomology rings if and only if they are algebra equivalent.

At the conference Randell announced a significant progress in the topological study of deformations of arrangements. He showed [117] that lattice-isotopic arrangements are topologically isomorphic.

6 The Cohomology of $M(\mathcal{A})$

In this section all arrangements are over the complex numbers. All homology and cohomology groups have integer coefficients.

The Poincaré Polynomial of the Complement

The first result of this section is due to Brieskorn [20]. His proof involves some algebraic geometry. Falk gave a topological argument in his thesis [40]. The results of Goresky and MacPherson [54] also provide a proof.

THEOREM 6.1. *Let $\mathcal{A}$ be an affine complex arrangement and let $M = M(\mathcal{A})$.*
(i) *For $X \in L(\mathcal{A})$ let $M_X = M(\mathcal{A}_X)$. For $k \geq 0$ there are isomorphisms*

$$\theta_k : \bigoplus_{X \in L_k} H^k(M_X) \to H^k(M)$$

induced by the inclusion maps $i_X : M \to M_X$.
(ii) *The groups $H^k(M)$ are free abelian.*

DEFINITION 6.2. *Let $\mathcal{A}$ be an affine complex arrangement with complement $M = M(\mathcal{A})$. Let $b_p(M) = \operatorname{rank} H^p(M)$ be the Betti numbers of M. The Poincaré polynomial of the complement is*

$$P(M(\mathcal{A}), t) = \sum_{p \geq 0} b_p(M) t^p.$$

THEOREM 6.3. *Let $\mathcal{A}$ be an affine complex arrangement with complement $M(\mathcal{A})$. Then*

$$P(M(\mathcal{A}), t) = \pi(\mathcal{A}, t).$$

Proof. Assume first that $\mathcal{A}$ is central. We argue by induction on $r(\mathcal{A})$. If $r(\mathcal{A}) = 0$ then both sides equal 1. If $r(\mathcal{A}) > 0$ then it follows from Proposition 5.3 that $M = M^* \times \mathbf{C}^*$ and hence its Euler characteristic $e(M) = 0$. Let $q = r(\mathcal{A})$ and $T = T(\mathcal{A})$. It suffices to show that

$$b_q(M) = (-1)^q \mu(T).$$

If $X \in L$ and $r(X) < q$ then we may assume by induction that $b_{r(X)}(M_X) = (-1)^{r(X)}\mu(X)$. Now use this together with Theorem 6.1 to get:

$$\begin{aligned} 0 &= \sum_{k=0}^{q} b_k(M) \\ &= \sum_{k=0}^{q-1}(-1)^k \sum_{X\in L_k} b_k(M_X) + (-1)^q b_q(M) \\ &= \sum_{r(X)<q} \mu(X) + (-1)^q b_q(M) \\ &= -\mu(T) + (-1)^q b_q(M). \end{aligned}$$

This proves the result for the central arrangement $\mathcal{A} = \mathcal{A}_T$. If $\mathcal{A}$ is an affine arrangement of rank q with maximal elements $T_1, \ldots, T_s$ then the argument above shows that $b_q(M_{T_j}) = (-1)^q \mu(T_j)$ for $1 \leq j \leq s$. The result follows for all affine arrangements of the same rank by Theorem 6.1.

COROLLARY 6.4. *Let $(\mathcal{A}_{\mathbf{R}}, V_{\mathbf{R}})$ be a real arrangement and let $(\mathcal{A}_{\mathbf{C}}, V_{\mathbf{C}})$ be its complexification. Let $M_{\mathbf{R}} = M(\mathcal{A}_{\mathbf{R}})$ and $M_{\mathbf{C}} = M(\mathcal{A}_{\mathbf{C}})$ be the real and complex complements. Let $b_i(M_{\mathbf{R}})$ and $b_i(M_{\mathbf{C}})$ be their respective Betti numbers. Then*

$$\sum_{i\geq 0} b_i(M_{\mathbf{R}}) = \sum_{i\geq 0} b_i(M_{\mathbf{C}}).$$

Proof. Since $M_{\mathbf{R}}$ is a disjoint union of chambers and each chamber is contractible, $b_i(M_{\mathbf{R}}) = 0$ for $i > 0$ and $b_0(M_{\mathbf{R}})$ is the number of chambers. By Theorem 2.32 $b_0(M_{\mathbf{R}}) = \pi(\mathcal{A}_{\mathbf{R}}, 1)$. By Theorem 6.3 we have

$$\sum_{i\geq 0} b_i(M_{\mathbf{C}}) = P(M(\mathcal{A}_{\mathbf{C}}), 1) = \pi(\mathcal{A}_{\mathbf{C}}, 1).$$

The result follows from the fact that $L(\mathcal{A}_{\mathbf{R}}) = L(\mathcal{A}_{\mathbf{C}})$.

Next we give a topological interpretation of restriction and deletion.

DEFINITION 6.5. *Let $(\mathcal{A}, \mathcal{A}', \mathcal{A}'')$ be a triple of arrangements with distinguished hyperplane $H_0 \in \mathcal{A}$. Let $M = M(\mathcal{A}), M' = M(\mathcal{A}'), M'' = M(\mathcal{A}'')$.*

LEMMA 6.6. *The spaces M, M' are open complex manifolds of complex dimension ℓ. The map $i : M \to M'$ is the inclusion of a submanifold.*

Proof. M' is the complement of an algebraic set in V and M is the complement of an algebraic set in M'.

LEMMA 6.7. (i) $M = M' - M''$,
(ii) $M'' = M' \cap H_0$,
(iii) *M'' is a submanifold of M' of complex codimension one.*

Proof. Assertions (i) and (ii) are clear. For (iii) note that H_0 has codimension one in V. The conclusion follows since M'' is open in H_0 and M' is open in V.

LEMMA 6.8. *If $H_0 \in \mathcal{A}$ then the submanifold $M'' \subset M'$ has a tubular neighborhood $E \subset M'$ which has the structure of a trivial* **C***-bundle over M''.*

Proof. View M' as an open smooth manifold of complex dimension ℓ and M'' a smooth submanifold of complex codimension 1. The existence of the tubular neighborhood is a general fact, see [66, p.110]. The bundle is trivial because it is a subbundle of a tubular neighborhood of H_0 in V, and the latter is clearly trivial.

Call the bundle $\xi = (E, M'', p)$ and view E as a subset of M' with inclusion map $q : E \to M'$. Let E_0 be the complement of the zero section in E. We may identify the zero section with M'' and $E_0 = E - M''$. By Lemma 6.7.i $M = M' - M''$ so we have

$$E \cap M = E \cap M' - E \cap M'' = E - M'' = E_0.$$

LEMMA 6.9. *Let $\xi = (E, M'', p)$ be a tubular neighborhood of M'' in M'. Then there are isomorphisms for $k \geq 1$:*

$$\tau : H^{k+1}(M', M) \to H^{k-1}(M'').$$

Proof. Since $E_0 = E \cap M$ the embedding $q : E \to M'$ is an inclusion of pairs $q : (E, E_0) \to (M', M)$. This inclusion is excision of the closed subset $M' - E \subset M$ and therefore q^* is an isomorphism. Since $(E, E_0) \approx M'' \times (\mathbf{C}, \mathbf{C}^*)$ we have isomorphisms $H^{k-1}(M'') \xrightarrow{p^*} H^{k-1}(E) \xrightarrow{r} H^{k+1}(E, E_0)$. Here r is the Thom isomorphism for the trivial bundle ξ. Then $\tau = (p^*)^{-1} r^{-1} q^*$.

THEOREM 6.10. *Let $\mathcal{A}$ be an affine complex arrangement with $M = M(\mathcal{A})$.*
(i) *The groups $H^k(M)$ are free abelian.*
(ii) *For $k \geq 0$ there exist split short exact sequences:*

$$0 \to H^k(M') \xrightarrow{i^*} H^k(M) \xrightarrow{\phi} H^{k-1}(M'') \to 0.$$

Proof. Part (i) follows from Theorem 6.1. To prove (ii) consider the long exact sequence of the pair (M', M) in cohomology:

$$\cdots \to H^k(M') \xrightarrow{i^*} H^k(M) \xrightarrow{\delta} H^{k+1}(M', M) \xrightarrow{j^*} H^{k+1}(M') \to \cdots.$$

We may use the isomorphism of Lemma 6.9 to replace $H^{k+1}(M', M)$ by $H^{k-1}(M'')$ and let $\phi = \tau\delta$ and $\psi = j^*\tau^{-1}$ to obtain the long exact sequence

$$\cdots \to H^k(M') \xrightarrow{i^*} H^k(M) \xrightarrow{\phi} H^{k-1}(M'') \xrightarrow{\psi} H^{k+1}(M') \to \cdots.$$

It follows from Theorem 6.3 and Corollary 2.29 that the Betti numbers satisfy $b_k(M) = b_k(M') + b_{k-1}(M'')$. The conclusion follows by induction on k.

PROPOSITION 6.11. Theorem 6.1 *is equivalent to* Theorem 6.10.

Proof. We derived Theorem 6.10 from Theorem 6.1. For the converse it suffices to prove (i) since (ii) is the same as 6.10.i. Recall that $\mathcal{A}_X = \{H \in \mathcal{A} | X \subset H\}$ and define $\mathcal{A}'_X = \mathcal{A}_X - \{H_0\}$ and $\mathcal{A}''_X = (\mathcal{A}_X)^{H_0}$. Let $M'_X = M(\mathcal{A}'_X)$ and $M''_X = M(\mathcal{A}''_X)$. We agree that if $X \not\subset H_0$ then $M'_X = M_X$ and $M''_X = \emptyset$. We use induction on $|\mathcal{A}|$. The assertion is clear for $|\mathcal{A}| = 1$. It follows from 6.10 that for every $X \in L$ and for $k \geq 0$ we have

$$H^k(M_X) \approx H^k(M'_X) \oplus H^{k-1}(M''_X).$$

Take the direct sum over $L_k = L_k(\mathcal{A})$. Let $L'_k = L_k(\mathcal{A}')$. We show that

$$\bigoplus_{X \in L_k} H^k(M'_X) = \bigoplus_{X \in L'_k} H^k(M'_X).$$

If $X \in L(\mathcal{A}')$ then $r'(X) = r(X)$. Thus it suffices to show that if $X \notin L(\mathcal{A}')$ then $H^k(M'_X) = 0$. If $X \notin L(\mathcal{A}')$ then $X = Y \cap H_0$ for some $Y \in L(\mathcal{A}')$ and $r(Y) = r(X) - 1 = k - 1$. Thus $\mathcal{A}'_X = \mathcal{A}_X - \{H_0\} = \mathcal{A}_Y$ and $M'_X = M_Y$. Since $|\mathcal{A}_Y| < |\mathcal{A}_X| \leq |\mathcal{A}|$ the induction hypothesis applies to $\mathcal{A}_Y$. Since $k > r(Y)$ this gives $H^k(M_Y) = 0$ and hence $H^k(M'_X) = H^k(M_Y) = 0$.

Next let $L''_{k-1} = L_{k-1}(\mathcal{A}'')$. We show that

$$\bigoplus_{X \in L_k} H^{k-1}(M''_X) = \bigoplus_{X \in L''_{k-1}} H^{k-1}(M''_X).$$

If $X \in L(\mathcal{A}'')$ then $r''(X) = r(X) - 1$ so the terms agree. If $X \notin L(\mathcal{A}'')$ then $H_0 \notin \mathcal{A}_X$ and $M'' = \emptyset$. The argument is completed using 6.10 again, and the induction hypothesis:

$$\begin{aligned} H^k(M) &\approx H^k(M') \oplus H^k(M'') \\ &\approx \left(\bigoplus_{X \in L'_k} H^k(M'_X) \right) \oplus \left(\bigoplus_{X \in L''_{k-1}} H^{k-1}(M''_X) \right) \\ &= \bigoplus_{X \in L_k} (H^k(M'_X) \oplus H^{k-1}(M''_X)) \\ &\approx \bigoplus_{X \in L_k} H^k(M_X). \end{aligned}$$

This completes the proof.

REMARK 6.12. *It would be interesting to find a direct argument for* Theorem 6.10.

The Cohomology Ring

Our next aim is to show that the integral cohomology ring $H^*(M)$ is generated by 1 and the elements of $H^1(M)$. It follows from Theorem 6.1 at $k = 1$ that there is an isomorphism

$$\theta_1 : \bigoplus_{H \in \mathcal{A}} H^1(M_H) \to H^1(M)$$

induced by the inclusion maps $i_H : M \to M_H$.

DEFINITION 6.13. *Let α_H be a linear form with kernel H. The map $\alpha_H : V \to \mathbf{C}$ restricts to $\alpha : M_H \to \mathbf{C}^*$. Choose the canonical generator of $H^1(\mathbf{C}^*)$ as dz/z and let*

$$\omega_H = \alpha_H^*(dz/z) = d\alpha_H/\alpha_H.$$

Let $\langle \omega_H \rangle$ be the cohomology class of ω_H in $H^1(M_H)$ and let $[\omega_H] = i^*\langle \omega_H \rangle$.

LEMMA 6.14. $H^1(M) = \bigoplus_{H \in \mathcal{A}} \mathbf{Z}[\omega_H]$.

Proof. This follows from Theorem 6.1 and the naturality of the maps.

LEMMA 6.15. *The natural orientation of the $\mathbf{R}^2$-bundle $\xi = (E, M'', p)$ has Thom class $u \in H^2(E, E_0)$ given by $u = q^*\delta[\omega_{H_0}]$.*

Proof. Write $M_0 = M_{H_0}, \alpha_0 = \alpha_{H_0}$ and $\omega_0 = \omega_{H_0}$. In the cohomology exact sequence of the pair (V, M_0) we have $H^1(V) = H^2(V) = 0$ so $\delta : H^1(M_0) \to H^2(V, M_0)$ is an isomorphism. Since $\langle \omega_0 \rangle$ generates $H^1(M_0)$ $\delta\langle \omega_0 \rangle$ generates $H^2(V, M_0)$. Let $x \in M''$ be any point and let $F = p^{-1}(x)$. Let $F_0 = F \cap E_0$. Let $k : (F, F_0) \to (V, M_0)$ be the inclusion of the fiber. Since $\alpha_0 : (F, F_0) \to (\mathbf{C}, \mathbf{C}^*)$ is a homotopy equivalence of pairs, k^* is an isomorphism. If ξ is given the natural orientation, then k^* sends a positive generator to a positive generator. We have the following inclusion of pairs with $k = jqi$:

$$(F, F_0) \xrightarrow{i} (E, E_0) \xrightarrow{q} (M', M) \xrightarrow{j} (V, M_0).$$

In cohomology we get the commutative diagram:

$$\begin{array}{ccccccc} H^2(V, M_0) & \xrightarrow{j^*} & H^2(M', M) & \xrightarrow{q^*} & H^2(E, E_0) & \xrightarrow{i^*} & H^2(F, F_0) \\ \uparrow \delta & & \uparrow \delta & & & & \\ H^1(M_0) & \xrightarrow{i_0^*} & H^1(M) & & & & \end{array}$$

Since $i^*q^*j^* = k^*$ is an isomorphism and $\delta\langle \omega_0 \rangle$ generates $H^2(V, M_0)$ it follows that $i^*q^*j^*\delta\langle \omega_0 \rangle = i^*q^*\delta[\omega_0]$ generates $H^2(F, F_0)$. Thus the restriction of $q^*\delta[\omega_0]$ is a generator of $H^2(F, F_0)$ for every fiber. This characterizes the Thom class u of the bundle ξ.

Our next aim is to prove that the classes $[\omega_H]$ are algebra generators for $H^*(M)$. Let $H \in \mathcal{A}'$ and let $i'_H : M' \to M_H$. Define $[\omega'_H] = (i'_H)^*\langle\omega_H\rangle$. Naturality implies that for $i : M \to M'$ we have $i^*[\omega'_H] = [\omega_H]$. Next suppose $B \in \mathcal{A}''$. Then $B = H_0 \cap H$ for some $H \in \mathcal{A}'$. Let $M''_B = H_0 - B$. Then we have $i''_B : M'' \to M''_B$. Let $\langle\omega''_B\rangle \in H^1(M''_B)$ be the canonical generator and let $[\omega''_B] = (i''_B)^*\langle\omega''_B\rangle$.

LEMMA 6.16. *Let $B \in \mathcal{A}''$. Then for every $H \in \mathcal{A}'$ with $B = H_0 \cap H$ we have $[\omega''_B] = z^*q^*[\omega'_H]$. Here $q : E \to M'$ is the tubular neighborhood and $z : M'' \to E$ its zero section.*

Proof. This follows from the equality of the maps

$$M'' \xrightarrow{z} E \xrightarrow{q} M' \xrightarrow{i'_H} M_H \xrightarrow{\alpha_H} \mathbf{C}^*,$$

$$M'' \xrightarrow{i''_B} M''_B \xrightarrow{\alpha_H|_{H_0}} \mathbf{C}^*$$

after we pass to cohomology.

LEMMA 6.17. *Let $H \in \mathcal{A}'$. Then for any $[a] \in H^k(M)$ we have*

$$\delta([a] \cup [\omega_H]) = \delta[a] \cup [\omega'_H].$$

Proof. Stability [35, p. 220] of the diagram

$$\begin{array}{ccccc} H^k(M) \otimes H^1(M') & \xrightarrow{\mathrm{id}\otimes i^*} & H^k(M) \otimes H^1(M) & \xrightarrow{\cup} & H^{k+1}(M) \\ \downarrow \delta \otimes \mathrm{id} & & & & \downarrow \delta \\ H^{k+1}(M', M) \otimes H^1(M') & \longrightarrow & \xrightarrow{\cup} & \longrightarrow & H^{k+2}(M', M) \end{array}$$

gives $\delta([a] \cup i^*[\omega'_H]) = \delta[a] \cup [\omega'_H]$ and we noted that $i^*[\omega'_H] = [\omega_H]$ by naturality.

LEMMA 6.18. *Let $B \in \mathcal{A}''$. Then for any $H \in \mathcal{A}'$ with $B = H_0 \cap H$ we have $\phi([\omega_0] \cup [\omega_H]) = [\omega''_B]$.*

Proof. Recall that $\phi = (p^*)^{-1}r^{-1}q^*\delta$. Thus we have

$$\begin{aligned} (p^*)^{-1}r^{-1}q^*\delta([\omega_0] \cup [\omega_H]) &= (p^*)^{-1}r^{-1}q^*(\delta[\omega_0] \cup [\omega'_H]) \\ &= (p^*)^{-1}r^{-1}(q^*\delta[\omega_0] \cup q^*[\omega'_H]) \\ &= (p^*)^{-1}r^{-1}(u \cup q^*[\omega'_H]) \\ &= (p^*)^{-1}q^*[\omega'_H] \\ &= z^*q^*[\omega'_H] \\ &= [\omega''_B]. \end{aligned}$$

Here we used naturality, the fact that cupping with u is r, and that p and z are homotopy inverses.

THEOREM 6.19. *The integral cohomology ring $H^*(M)$ is generated by 1 and the classes $[\omega_H]$, for $H \in \mathcal{A}$.*

Proof. We use induction on $|\mathcal{A}|$. For $|\mathcal{A}| = 1$ the assertion follows from the residue theorem. For $|\mathcal{A}| > 1$ we use the direct sum decomposition

$$H^*(M) \approx H^*(M') \oplus H^*(M'')$$

which follows from Theorem 6.10, together with the induction hypothesis. Thus we may assume that $H^*(M')$ is generated by $[\omega'_H]$ for $H \in \mathcal{A}'$, and $H^*(M'')$ is generated by $[\omega''_B]$ for $B \in \mathcal{A}''$. We noted that $i^*[\omega'_H] = [\omega_H]$ and Lemma 6.17 shows that $[\omega''_B]$ is a product of elements of the required form.

COROLLARY 6.20. *The* **C**-*algebra $H^*(M;\mathbf{C})$ is generated by 1 and the classes $[\omega_H]$ for $H \in \mathcal{A}$.*

In the next section we shall define an algebra $R = R(\mathcal{A})$ over an arbitrary field **K** generated by 1 and ω_H for $H \in \mathcal{A}$. We will show that the Poincaré polynomial of $R(\mathcal{A})$ is

$$P(R(\mathcal{A}), t) = \pi(\mathcal{A}, t).$$

Thus we obtain a result of Brieskorn [20]:

THEOREM 6.21. *The surjective map $\omega_H \to [\omega_H]$ induces an isomorphism of graded algebras $R(\mathcal{A}) \approx H^*(M(\mathcal{A});\mathbf{C})$.*

This shows that there are no relations in cohomology other than what is imposed by the algebraic relations. The main result of the next section is that there is an isomorphism of algebras $A(\mathcal{A}) \approx R(\mathcal{A})$ which sends a_H to ω_H. We may apply this result when $\mathbf{K} = \mathbf{C}$ to obtain a structure theorem for $H^*(M;\mathbf{C})$ as follows.

THEOREM 6.22. *Let $\mathcal{A}$ be an arrangement over* **C**, *let $M = M(\mathcal{A})$, and let $A = A(\mathcal{A})$. The map $a_H \to [\omega_H]$ induces an isomorphism $A \to H^*(M, \mathbf{C})$ of graded* **C**-*algebras.*

This result gives a presentation of $H^*(M;\mathbf{C})$ in terms of generators and the relation ideal. Note that the result is only proved for central arrangements because $A(\mathcal{A})$ was only defined for central arrangements. We have no similar description for affine arrangements.

Alexander Duality

The algebra $A(\mathcal{A})$ describes the cohomology of $M(\mathcal{A})$. Its classes are "torical" in the sense that they are products of one-dimensional classes. The elements of the algebra $B(\mathcal{A})$ are "spherical" in the sense of Theorem 4.21. It is natural to ask whether these classes also have a topological interpretation.

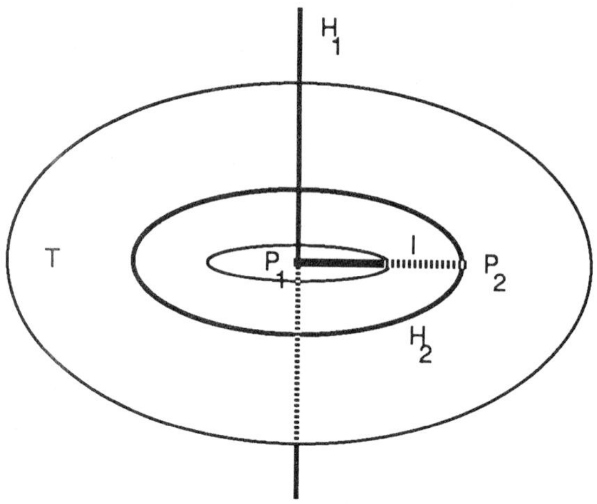

Figure 19: Falk's Linking

Consider the unit sphere $S = S^{2\ell-1} \subset V$ and let $\hat{M} = S \cap M$, $\hat{N} = S \cap N$. Clearly $\hat{M}$ is a strong deformation retract of M. Alexander duality [133, p.296] in the compact $(2\ell - 1)$-manifold S for the compact polyhedron $\hat{N}$ gives:

$$H_{q+1}(S, \hat{M}) \approx H^{2\ell-q-2}(\hat{N}).$$

Thus for $1 \le q \le 2\ell - 3$ we have

$$H_q(\hat{M}) \approx H^{2\ell-q-2}(\hat{N}).$$

This gives rise to a linking pairing in S. In his thesis [40] Falk interpreted this as a geometric linking. He constructed an embedding of the complex $F(\mathcal{A})$ in $\hat{N}$ which induces isomorphism in homology and in homotopy through dimension $(\ell - 2)$. It follows from Theorems 4.20 and 4.21 that this is a geometric representation of $\mathcal{H}_\ell(\mathcal{A}) = B_\ell(\mathcal{A})$. He also constructed embeddings of ℓ-tori in $\hat{M}$ which represent the ℓ-dimensional homology of $\hat{M}$. Then he showed that these classes link appropriately.

THEOREM 6.23. *If $S = (H_1, \ldots H_\ell)$ is independent, so the cycle*

$$z_S = \sum_{\pi \in \mathrm{Sym}(\ell)} (-1)^{\ell-1}(\mathrm{sign}\,\pi)(H_{\pi 1}, H_{\pi 1} \cap H_{\pi 2}, \ldots, H_{\pi 1} \cap \cdots \cap H_{\pi(\ell-1)})$$

is a generator of $\tilde{H}_{\ell-2}(\hat{N})$, and $\tau_S \in H_\ell(\hat{M})$ is the algebraic dual of $\omega_S \in H^\ell(\hat{M})$ then the linking number of z_S and τ_S is ± 1.

The case $\ell = 2$ is illustrated in Figure 19. Let $H_i = \ker x_i$ for $i = 1, 2$. Stereographic projection from $(0, 1)$ sends S^3 onto R^3. The image of $\hat{N}$ consists of the vertical axis and the unit circle in the horizontal plane. The embedded torus

$$T = \{(x_1, x_2) \in S^3 \,|\, |x_1| = 1, |x_2| = 1\}$$

is a generator of $H_2(\hat{M}) \approx \mathbf{Z}$. The Folkman complex consists of one point for each hyperplane, $F(\mathcal{A}) = \{P_1, P_2\}$. Embed $F(\mathcal{A}) \to \hat{N}$ by letting $P_i \in H_i \cap S^3$. The group $\tilde{H}_0(\hat{N}) \approx \mathbf{Z}$ is generated by the cycle $z = (P_2 - P_1)$. The fact that T and z have linking number ± 1 follows because we can choose an embedded 1-chain I with $\partial I = z$ such that the geometric intersection $I \cap T$ is a single point.

Alexander duality in the open 2ℓ-manifold $V = \mathbf{C}^\ell = \mathbf{R}^{2\ell}$ for the closed polyhedron N and its complement M gives for $1 \leq q \leq 2\ell - 1$:

$$H_q(M) \approx H_c^{2\ell - q - 1}(N)$$

where H_c denotes cohomology with compact supports. It would be interesting to give a geometric interpretation of Alexander duality for all q.

Arrangements of Subspaces

In a recent book [54] Goresky and MacPherson considered real arrangements of affine subspaces of possibly various dimensions. We shall call these **subspace** arrangements. In [54, pp. 237-244] they computed the groups $H_*(M(\mathcal{A}); \mathbf{Z})$ using the order complex of Definition 4.2. They partially order $L(\mathcal{A})$ by inclusion. The statement of their theorem given below is adjusted for this difference of conventions.

DEFINITION 6.24. *Let $(\mathcal{A}, V)$ be a real arrangement of subspaces. Let $L = L(\mathcal{A})$ be the set of all intersections partially ordered by reverse inclusion with rank function $r(X) = \operatorname{codim} X$. Given $X, Y \in L$ with $X < Y$ define the segments:*

$$L_{(X,Y)} = \{Z \in L | X < Z < Y\},$$
$$L_{<X} = \{Z \in L | Z < X\}.$$

Write $KL_* = K(L_*)$ for the corresponding order complex.

THEOREM 6.25. *The homology of the complement $M = M(\mathcal{A})$ is given by*

$$H_i(M; \mathbf{Z}) = \bigoplus_{X \in L} H^{r(X)-i-1}(KL_{<X}, KL_{(V,X)}; \mathbf{Z})$$

with the convention that $H^{-1}(\emptyset, \emptyset) = \mathbf{Z}$. Thus V contributes a copy of $\mathbf{Z}$ to $H_0(M)$.

EXAMPLE 6.26. *Two lines in $\mathbf{R}^3$.*

Let x, y, z be the usual coordinates in $V = \mathbf{R}^3$. Let H be the x-axis, let K be the y-axis, and let $\mathcal{A} = \{H, K\}$. The columns of the table below list the elements of $L(\mathcal{A})$, their rank, the complexes $KL_{<X}$, $KL_{(V,X)}$, and their contributions to the Betti numbers. Here $\{P\}$ denotes a point labeled P and $[H, K]$ denotes an interval with end points H and K.

X	$r(X)$	$KL_{<X}$	$KL_{(V,X)}$	b_0	b_1
V	0	$\emptyset$	$\emptyset$	1	
H	2	$\{V\}$	$\emptyset$		1
K	2	$\{V\}$	$\emptyset$		1
0	3	$[H,K]$	$\{H,K\}$		1

It follows that $P(M(\mathcal{A}),t) = 1+3t$. To see this directly, retract M radially onto the unit sphere minus the four points $(\pm 1,0,0), (0,\pm 1,0)$. Note that the lines in $\mathcal{A}$ are not in general position. If $\mathcal{B}$ is an arrangement in $\mathbf{R}^3$ consisting of two lines in general position, then it is easy to check that $P(M(\mathcal{B}),t) = 1+2t$. The two circles linking the lines in $\mathcal{B}$ generate $H_1(M(\mathcal{B}))$ but the two circles which link H and K in $M(\mathcal{A})$ do not generate $H_1(M(\mathcal{A}))$. Thus the analog of Theorem 6.19 fails for subspace arrangements.

7 Differential Forms

In this section we study the algebra of differential forms generated by 1 and the $\omega_H = d\alpha_H/\alpha_H$ defined in the last section. This algebra was first computed by Arnold [3] for the braid arrangement. Brieskorn [20] defined it for all arrangements and showed that it is isomorphic to the cohomology algebra. Its isomorphism with $A(\mathcal{A})$ was established in [100]. In all these topological considerations the field was $\mathbf{C}$. Our presentation is based on [109], where the properties of $R(\mathcal{A})$ over an arbitrary field $\mathbf{K}$ were first studied. It is important to note that the definition of $R(\mathcal{A})$ involves the linear forms α_H, and thus this algebra is not obviously a combinatorial invariant of $\mathcal{A}$. Its combinatorial nature is a consequence of the main theorem of this section which establishes an algebra isomorphism between $A(\mathcal{A})$ and $R(\mathcal{A})$.

The de Rham Complex

Let $(\mathcal{A}, V)$ be an arrangement over $\mathbf{K}$. Let S be the symmetric algebra of V^* and let F be the quotient field of S. It will sometimes be convenient to indicate the dependence of S and F on V. In this case we write $S = \mathbf{K}[V]$ and $F = \mathbf{K}(V)$. We view $F \otimes_{\mathbf{K}} V^*$ as a vector space over F by defining $f(g \otimes \alpha) = fg \otimes \alpha$ where $f, g \in F$ and $\alpha \in V^*$. There exists a unique $\mathbf{K}$-linear map $d : F \to F \otimes V^*$ such that $d(fg) = f(dg) + g(df)$ for $f, g \in F$ and $d\alpha \in \mathbf{K}$ for $\alpha \in V^*$. Recall that we have chosen a basis $x_1, \dots, x_\ell$ for V^* so we may identify the symmetric algebra of V^* with the polynomial algebra $S = \mathbf{K}[x_1, \dots, x_\ell]$ and its quotient field with the field of rational functions $F = \mathbf{K}(x_1, \dots, x_\ell)$. In terms of this basis the differential df is given by the usual formula

$$df = \sum_{i=1}^{\ell} \frac{\partial f}{\partial x_i} \otimes x_i = \sum_{i=1}^{\ell} \frac{\partial f}{\partial x_i} dx_i.$$

Note that $F \otimes V^* = F dx_1 \oplus \cdots \oplus F dx_\ell$.

DEFINITION 7.1. *Let $\Omega(V)$ be the exterior algebra of the F-vector space $F \otimes V^*$ graded by $\Omega(V) = \bigoplus_{p=0}^{\ell} \Omega^p(V)$ where*

$$\Omega^p(V) = \bigoplus_{1 \le i_1 < \cdots < i_p \le \ell} F dx_{i_1} \wedge \cdots \wedge dx_{i_p}.$$

For simplicity of notation we shall write $\omega\eta = \omega \wedge \eta$, for ω, $\eta \in \Omega(V)$. In particular we write $dx_1 \cdots dx_p = dx_1 \wedge \cdots \wedge dx_p$. We identify Ω^0 with F. The elements of $\Omega^p(V)$ are called rational differential p-forms on V. We list some well known properties of d for future reference.

PROPOSITION 7.2. *The map $d : F \to F \otimes V^*$ may be extended in a unique way to a $\mathbf{K}$-linear map $d : \Omega(V) \to \Omega(V)$ with the following properties:*
(i) $d^2 = 0$,
(ii) *if $\omega \in \Omega^p(V)$ and $\eta \in \Omega(V)$ then*

$$d(\omega\eta) = (d\omega)\eta + (-1)^p \omega(d\eta),$$

(iii) *if $\omega = \sum f_{i_1 \cdots i_p} dx_{i_1} \cdots dx_{i_p}$ where $1 \leq i_1 < \cdots < i_p \leq \ell$ and $f_{i_1 \cdots i_p} \in F$ then*

$$d\omega = \sum_{j=1}^{\ell} \sum (\partial f_{i_1 \cdots i_p} / \partial x_j) dx_j dx_{i_1} \cdots dx_{i_p}.$$

DEFINITION 7.3. *Let $\mathcal{A}$ be an arrangement. For $H \in \mathcal{A}$ let $\alpha_H \in V^*$ with $H = \ker \alpha_H$ and let $\omega_H = d\alpha_H / \alpha_H \in \Omega^1(V)$. Let $R = R(\mathcal{A})$ be the $\mathbf{K}$-subalgebra of $\Omega(V)$ generated by 1 and ω_H for $H \in \mathcal{A}$.*

Let $R_p = R \cap \Omega^p(V)$. Since R is generated by the homogeneous elements ω_H, it is naturally graded $R = \bigoplus_{p=0}^{\ell} R_p$.

To give the reader some intuitive idea why this algebra is again isomorphic to $A(\mathcal{A})$ we work out the analog of Examples 3.9 and 3.32.

EXAMPLE 7.4. *Suppose $\ell = 2$ and $\mathcal{A} = \{H_1, \ldots, H_n\}$. Write $\omega_i = \omega_{H_i}$. Then*

$$R(\mathcal{A}) = \mathbf{K} \oplus \bigoplus_{p=1}^{n} \mathbf{K}\omega_p \oplus \bigoplus_{k=1}^{n-1} \omega_k \omega_n.$$

We know that $R_0 = \mathbf{K}$ and that $R_p = 0$ for $p > 2$. By definition $\omega_1, \ldots, \omega_n$ span R_1 over $\mathbf{K}$. These 1-forms are linearly independent over $\mathbf{K}$ because the rational functions $1/\alpha_1, \ldots, 1/\alpha_n$ are linearly independent over $\mathbf{K}$. Since $\omega_i^2 = 0$ and $\omega_i\omega_j = -\omega_j\omega_i$, the space R_2 is spanned over $\mathbf{K}$ by the $\omega_i\omega_j$ with $i < j$. In order to discover the remaining relations among these generators let x, y be a basis for V^* and write $\alpha_i = a_i x + b_i y$ with a_i, $b_i \in \mathbf{K}$. Then $\omega_i = (a_i/\alpha_i)dx + (b_i/\alpha_i)dy$ and we have

$$d\alpha_i d\alpha_j = (a_i b_j - b_i a_j) dx dy.$$

Thus for any i, j, k we have

$$\begin{aligned} &\alpha_k d\alpha_i d\alpha_j + \alpha_i d\alpha_j d\alpha_k + \alpha_j d\alpha_k d\alpha_i \\ &= \det \begin{bmatrix} a_i & a_j & a_k \\ b_i & b_j & b_k \\ \alpha_i & \alpha_j & \alpha_k \end{bmatrix} dx dy = 0 \end{aligned}$$

because the third row is a linear combination of the first two. If we multiply this equation by $1/(\alpha_i\alpha_j\alpha_k)$ we get:

$$\omega_i\omega_j + \omega_j\omega_k + \omega_k\omega_i = 0.$$

In particular we have $\omega_i\omega_j = \omega_i\omega_n - \omega_j\omega_n$ if $1 \le i < j \le n$ so R_2 is spanned by the elements $\omega_k\omega_n$ for $1 \le k < n$. It remains to show that these elements are linearly independent over $\mathbf{K}$. Define an F-linear map $\partial : \Omega^2(V) \to \Omega^1(V)$ by $\partial(f dx dy) = f x dy - f y dx$. Then $\partial(\omega_i\omega_j) = \omega_j - \omega_i$. If $\sum_{k=1}^{n-1} c_k\omega_k\omega_n = 0$ with $c_k \in \mathbf{K}$ then applying ∂ gives $\sum_{k=1}^{n-1} c_k(\omega_n - \omega_k) = 0$. Since $\omega_1, \dots, \omega_n$ are linearly independent over $\mathbf{K}$, we get $c_1 = \dots = c_{n-1} = 0$. This proves the assertion.

LEMMA 7.5. *There exists a surjective homomorphism* $\gamma : A(\mathcal{A}) \to R(\mathcal{A})$ *of graded* $\mathbf{K}$*-algebras such that* $\gamma(a_H) = \omega_H$ *for all* $H \in \mathcal{A}$.

Proof. Define a $\mathbf{K}$-algebra homomorphism $\nu : E \to R$ by $\nu(e_H) = \omega_H$. To prove that ν induces a homomorphism $\gamma : A \to R$ we must show that $\nu(I) = 0$. Thus we need to show that if $S = (H_1, \dots, H_p)$ is dependent then $\nu(\partial e_S) = 0$. First note that $\nu(e_S) = 0$ because if $(H_1, \dots, H_p)$ is dependent then $d\alpha_{H_1} \cdots d\alpha_{H_p} = 0$ and therefore $\omega_{H_1} \cdots \omega_{H_p} = 0$. We prove that $\nu(\partial e_S) = 0$ by induction on p. Write $\alpha_i = \alpha_{H_i}$ and $\omega_i = \omega_{H_i}$. For $p = 2$ if S is dependent then $S = (H, H)$ for some $H \in \mathcal{A}$ so $\partial e_S = 0$ and the assertion is clear. As before, write $S_k = (H_1, \dots, \hat{H}_k, \dots, H_p)$ for $k = 1, \dots, p$. If S_k is dependent for some k then $\nu(e_{S_k}) = 0$. We may assume by induction that $\nu(\partial e_{S_k}) = 0$. Since $e_S = (-1)^{k-1} e_{H_k} e_{S_k}$ it follows from Lemma 3.4.ii that $(-1)^{k-1}\partial(e_S) = \partial(e_{H_k} e_{S_k}) = e_{S_k} - e_{H_k}\partial(e_{S_k})$ so $\nu(\partial e_S) = 0$ and we are done. Thus we may assume that every S_k is independent. Equivalently, no proper subset of $\alpha_1, \dots, \alpha_p$ is linearly dependent. Since $\alpha_1, \dots, \alpha_p$ is a linearly dependent set, there exist $c_i \in \mathbf{K}$, all nonzero, with $\sum_{i=1}^{p} c_i\alpha_i = 0$. If we replace α_i by $c_i\alpha_i$ then ω_i is unchanged. Thus we may assume that $\sum \alpha_i = 0$. Suppose $1 \le j \le p - 1$. Then $\sum d\alpha_i = 0$ implies

$$\left(\sum d\alpha_i\right)(d\alpha_1 \cdots \widehat{d\alpha_j}\widehat{d\alpha_{j+1}} \cdots d\alpha_p) = 0.$$

Thus

$$d\alpha_1 \cdots \widehat{d\alpha_{j+1}} \cdots d\alpha_p = -d\alpha_1 \cdots \widehat{d\alpha_j} \cdots d\alpha_p.$$

Define η_j by $\alpha_j\eta_j = (-1)^{j-1}\omega_1 \cdots \widehat{\omega_j} \cdots \omega_p$. Then

$$\begin{aligned} \alpha_1 \cdots \alpha_p \eta_{j+1} &= (-1)^j d\alpha_1 \cdots \widehat{d\alpha_{j+1}} \cdots d\alpha_p \\ &= (-1)^{j-1} d\alpha_1 \cdots \widehat{d\alpha_j} \cdots d\alpha_p \\ &= \alpha_1 \cdots \alpha_p \eta_j. \end{aligned}$$

Let η be the common value of the η_j. Then

$$\begin{aligned}\nu(\partial e_S) &= \sum_{i=1}^{p}(-1)^{i-1}\omega_1\cdots\widehat{\omega_i}\cdots\omega_p \\ &= \sum_{i=1}^{p}\alpha_i\eta_i \\ &= \left(\sum_{i=1}^{p}\alpha_i\right)\eta \\ &= 0.\end{aligned}$$

Thus $\nu(I)=0$ and ν induces a surjective map $\gamma : A \to R$ such that $\gamma(a_H) = \omega_H$.

The Leray Residue

Our next aim is to study the properties of R with respect to deletion and restriction. Let $(\mathcal{A}, \mathcal{A}', \mathcal{A}'')$ be a triple of arrangements with distinguished hyperplane $H_0 \in \mathcal{A}$. We will construct $\mathbf{K}$-linear maps $i : R(\mathcal{A}') \to R(\mathcal{A})$ and $j : R(\mathcal{A}) \to R(\mathcal{A}'')$ and prove that there is an exact sequence

$$0 \to R(\mathcal{A}') \xrightarrow{i} R(\mathcal{A}) \xrightarrow{j} R(\mathcal{A}'') \to 0.$$

This and the corresponding exact sequence for $A(\mathcal{A})$ and the map γ will allow us to prove the isomorphism $A(\mathcal{A}) \approx R(\mathcal{A})$ by induction.

Note that $R(\mathcal{A}')$ and $R(\mathcal{A})$ are both subalgebras of $\Omega(V)$ and that $R(\mathcal{A}') \subset R(\mathcal{A})$. We define j with the help of the Leray residue map on differential forms. This definition is analogous to Pham's definition [112, Chapter III] in case $\mathbf{K} = \mathbf{C}$ and the forms are holomorphic. Let $\alpha_0 = \alpha_{H_0}$ and let S_0 be the localization of S at α_0. By definition S_0 is the subring of F consisting of all f/g such that $f, g \in S$ and g is prime to α_0. Let $\rho : V^* \to H_0^*$ be the restriction map and let $y_i = \rho(x_i)$. We may extend ρ uniquely to a $\mathbf{K}$-algebra homomorphism $\rho : S_0 \to \mathbf{K}(H_0)$. Both existence and uniqueness follow from the formula

$$\rho(f/g) = f(y_1, \ldots, y_\ell)/g(y_1, \ldots, y_\ell).$$

Note that $g(y_1, \ldots, y_\ell) \neq 0$ because g is prime to α_0. Define a $\mathbf{K}$-subalgebra Ω_0 of $\Omega(V)$ by

$$\Omega_0 = \bigoplus_{p=0}^{\ell} \bigoplus_{i_1<\cdots<i_p} S_0 dx_{i_1}\cdots dx_{i_p}.$$

This subalgebra does not depend on the basis for V^*.

LEMMA 7.6. *The map $\rho : S_0 \to \mathbf{K}(H_0)$ may be extended in a unique way to a $\mathbf{K}$-linear map $\rho : \Omega_0 \to \Omega(H_0)$ such that for ω, $\eta \in \Omega_0$, $f \in S_0$ and $\alpha \in V^*$ we have:*

(i) $\rho(\omega\eta) = \rho(\omega)\rho(\eta)$,
(ii) $\rho(f\omega) = \rho(f)\rho(\omega)$,
(iii) $\rho(d\alpha) = d\rho(\alpha)$,
(iv) *if* $\omega = \sum f_{i_1 \ldots i_p} dx_{i_1} \cdots dx_{i_p}$ *then*

$$\rho(\omega) = \sum f_{i_1 \cdots i_p}(y_1, \ldots, y_\ell) dy_{i_1} \cdots dy_{i_p}.$$

Proof. If $\omega = \sum f_{i_1 \cdots i_p} dx_{i_1} \cdots dx_{i_p}$ and ρ has the properties (i)–(iii) then

$$\begin{aligned} \rho(\omega) &= \sum \rho(f_{i_1 \cdots i_p}) \rho(dx_{i_1}) \cdots \rho(dx_{i_p}) \\ &= \sum \rho(f_{i_1 \cdots i_p}) dy_{i_1} \cdots dy_{i_p}. \end{aligned}$$

This shows that $\rho(\omega)$ is given by (iv) and proves uniqueness. To prove existence define $\rho(\omega)$ by (iv) and then (i)–(iii) are clear.

LEMMA 7.7. *Suppose $\alpha \in V^*$ and $\alpha \neq 0$. If $\omega \in \Omega_0$ and $(d\alpha)\omega = 0$ then there exists $\psi \in \Omega_0$ with $\omega = (d\alpha)\psi$.*

Proof. Choose a basis $x_1, \ldots, x_\ell$ for V^* such that $\alpha = x_1$. We may assume that ω is a p-form. Write $\omega = \sum f_{i_1 \cdots i_p} dx_{i_1} \cdots dx_{i_p}$ where $f_{i_1 \cdots i_p} \in S_0$ and the sum is over all $1 \leq i_1 < \cdots < i_p \leq \ell$. Then

$$0 = (dx_1)\omega = \sum f_{i_1 \cdots i_p} dx_1 dx_{i_1} \cdots dx_{i_p}$$

where the sum is over all $2 \leq i_1 < \cdots < i_p \leq \ell$. Thus $f_{i_1 \cdots i_p} = 0$ if $i_1 \geq 2$.

DEFINITION 7.8. *Say that $\phi \in \Omega(V)$ has at most a simple pole along H_0 if $\alpha_0 \phi \in \Omega_0$.*

LEMMA 7.9. *Suppose $\phi \in \Omega(V)$ has at most a simple pole along H_0 and that $d\phi = 0$. Then there exist ψ, $\theta \in \Omega_0$ such that*

$$\phi = (d\alpha_0/\alpha_0)\psi + \theta.$$

The form $\rho(\psi) \in \Omega(H_0)$ is uniquely determined by ϕ.

Proof. For simplicity write $\alpha = \alpha_0$. Since $d\phi = 0$ it follows from Lemma 7.2.ii that $d(\alpha\phi) = (d\alpha)\phi - \alpha(d\phi) = (d\alpha)\phi$. Since $\alpha\phi \in \Omega_0$ by hypothesis, and $d\Omega_0 \subset \Omega_0$ it follows from Lemma 7.7 that there exists $\theta \in \Omega_0$ such that $d(\alpha\phi) = (d\alpha)\theta$. Thus $(d\alpha)\phi = (d\alpha)\theta$, which implies $(d\alpha)\alpha(\phi - \theta) = 0$. Since $\alpha(\phi - \theta) \in \Omega_0$ it follows from Lemma 7.7 that there exists $\psi \in \Omega_0$ such that $\alpha(\phi - \theta) = (d\alpha)\psi$. This proves the existence of θ and ψ.

To prove the uniqueness of $\rho(\psi)$ it suffices to show that if $\psi, \theta \in \Omega_0$ and $(d\alpha/\alpha)\psi + \theta = 0$ then $\rho(\psi) = 0$. First note that $(d\alpha)\theta = 0$. It follows from Lemma 7.7 that there exists $\theta' \in \Omega_0$ such that $\theta = (d\alpha)\theta'$. Now $(d\alpha)(\psi+\alpha\theta') = (d\alpha)\psi+\alpha\theta = 0$. Since $\psi+\alpha\theta' \in \Omega_0$ we may apply Lemma 7.7 again to conclude that there exists $\theta'' \in \Omega_0$ with $\psi + \alpha\theta' = (d\alpha)\theta''$. Since $\rho(\alpha) = 0$ it follows from Lemma 7.6 that $\rho(\alpha\theta') = 0$ and $\rho((d\alpha)\theta'') = 0$. Thus $\rho(\psi) = 0$.

DEFINITION 7.10. *The uniquely determined form $\rho(\psi)$ is called the* **residue** *of ϕ along H_0 and we denote it* $\operatorname{res}(\phi)$.

If $H \in \mathcal{A}$ then

$$d\omega_H = d(d\alpha_H/\alpha_H) = (1/\alpha_H)d(d\alpha_H) - (1/\alpha_H^2)(d\alpha_H)(d\alpha_H) = 0$$

so $d(\omega_{H_1} \cdots \omega_{H_p}) = 0$ for all $H_1, \ldots, H_p \in \mathcal{A}$. Thus $d\phi = 0$ for all $\phi \in R(\mathcal{A})$. It is clear from the definition that each $\phi \in R(\mathcal{A})$ has at most a simple pole along H_0. Thus $\operatorname{res}(\phi)$ is defined for all $\phi \in R(\mathcal{A})$.

LEMMA 7.11. *Suppose $H_1, \ldots, H_p \in \mathcal{A}'$. Then*
(i) $\operatorname{res}(\omega_{H_1} \cdots \omega_{H_p}) = 0$,
(ii) $\operatorname{res}(\omega_{H_0}\omega_{H_1} \cdots \omega_{H_p}) = \omega_{H_0\cap H_1} \cdots \omega_{H_0\cap H_p}$,
(iii) $\operatorname{res}R(\mathcal{A}) \subset R(\mathcal{A}'')$.

Proof. In case $p = 0$ formulas (i) and (ii) are interpreted as $\operatorname{res}(1) = 0$ and $\operatorname{res}(\omega_{H_0}) = 1$. Let $\phi = \omega_{H_1} \cdots \omega_{H_p}$. We may choose $\psi = 0$ and $\theta = \phi$ in Lemma 7.9. This shows that $\operatorname{res}(\phi) = 0$ and proves (i). Now let $\phi = \omega_{H_0}\omega_{H_1} \cdots \omega_{H_p}$. We may choose $\psi = \omega_{H_1} \cdots \omega_{H_p}$ and $\theta = 0$ in Lemma 7.9. This shows that $\operatorname{res}(\phi) = \rho(\omega_{H_1} \cdots \omega_{H_p})$. By Lemma 7.6.i we have $\rho(\omega_{H_1} \cdots \omega_{H_p}) = \rho(\omega_{H_1}) \cdots \rho(\omega_{H_p})$. It remains to show that $\rho(\omega_{H_i}) = \omega_{H_0\cap H_i}$. If $H \in \mathcal{A}'$ then it follows from Lemma 7.6.i and iii that $\rho(\omega_H) = \rho(d\alpha_H/\alpha_H) = d\rho(\alpha_H)/\rho(\alpha_H)$. Since $\rho(\alpha_H)$ is a linear form on H_0 which defines the hyperplane $H_0 \cap H \in \mathcal{A}''$ we have $\rho(\omega_H) = \omega_{H_0\cap H}$. This proves (ii). To prove (iii) note that since $\omega_{H_0}^2 = 0$ it follows from the definition of $R(\mathcal{A})$ and $R(\mathcal{A}')$ that $R(\mathcal{A}) = R(\mathcal{A}') + \omega_{H_0}R(\mathcal{A}')$. Thus (iii) follows from (i) and (ii).

THEOREM 7.12. *Let $\mathcal{A}$ be an arrangement and let $R(\mathcal{A})$ be the algebra of differential forms generated by 1 and $\omega_H = d\alpha_H/\alpha_H$. The map $\gamma : A(\mathcal{A}) \to R(\mathcal{A})$ induces an isomorphism of graded* **K***-algebras such that $\gamma(a_H) = \omega_H$.*

THEOREM 7.13. *Let $(\mathcal{A}, \mathcal{A}', \mathcal{A}'')$ be a triple of arrangements with respect to $H_0 \in \mathcal{A}$. Let $i : R(\mathcal{A}') \to R(\mathcal{A})$ be the inclusion map and define $j : R(\mathcal{A}) \to R(\mathcal{A}'')$ by $j(\phi) = \operatorname{res}(\phi)$ for $\phi \in R(\mathcal{A})$, where $\operatorname{res}(\phi)$ is the residue of ϕ along H_0. Then there is an exact sequence:*

$$0 \to R(\mathcal{A}') \xrightarrow{i} R(\mathcal{A}) \xrightarrow{j} R(\mathcal{A}'') \to 0.$$

Proof. We prove Theorems 7.12 and 7.13 simultaneously by induction on $|\mathcal{A}|$. If $\mathcal{A}$ is empty then $A(\mathcal{A}) = \mathbf{K} = R(\mathcal{A})$ and the first result holds. The second assumes that $\mathcal{A}$ is nonempty. If $|\mathcal{A}| = 1$ then $\mathcal{A}'$ and $\mathcal{A}''$ are empty arrangements. Let $\mathcal{A} = \{H\}$. Then $R(\mathcal{A}) = \mathbf{K}+\mathbf{K}\omega_H$ and $R(\mathcal{A}') = \mathbf{K} = R(\mathcal{A}'')$ so both statements are clear. If $|\mathcal{A}| > 1$ then we see from Lemma 7.11.iii that $jR(\mathcal{A}) \subset R(\mathcal{A}'')$ and from Lemma 7.11.ii that j is surjective. It follows from Lemma 7.11.i that $ji = 0$ so $\operatorname{im}(i) \subset \ker(j)$. To prove that $\ker(j) \subset \operatorname{im}(i)$ consider the following diagram:

$$\begin{array}{ccccccccc}
0 & \rightarrow & A(\mathcal{A}') & \xrightarrow{i_A} & A(\mathcal{A}) & \xrightarrow{j_A} & A(\mathcal{A}'') & \rightarrow & 0 \\
 & & \gamma' \downarrow & & \gamma \downarrow & & \gamma'' \downarrow & & \\
0 & \rightarrow & R(\mathcal{A}') & \xrightarrow{i} & R(\mathcal{A}) & \xrightarrow{j} & R(\mathcal{A}'') & \rightarrow & 0
\end{array}$$

The diagram is commutative. This is clear for the left square by definition of i_A and i. For the right square it follows from Lemma 7.11. The top row is exact by Theorem 3.22. We may assume by the induction hypothesis in Theorem 7.12 that γ' and γ'' are isomorphisms. A diagram chase shows that $\ker(j) \subset \operatorname{im}(i)$. This proves that the second row of the diagram is exact. Thus Theorem 7.13 holds for $\mathcal{A}$. It follows from the Five Lemma that γ is an isomorphism, so Theorem 7.12 is also established for $\mathcal{A}$.

COROLLARY 7.14. *Let $\mathcal{A}$ be an arrangement and let $R(\mathcal{A})$ be be the algebra of differential forms generated by 1 and $\omega_H = d\alpha_H/\alpha_H$. The Poincaré polynomial of $R(\mathcal{A})$ is:*

$$P(R(\mathcal{A}), t) = \pi(\mathcal{A}, t).$$

We showed in Lemma 3.13 that ∂_A is a $\mathbf{K}$-derivation of A of degree -1. Since $\gamma : A \rightarrow R$ is an isomorphism of graded $\mathbf{K}$-algebras, it follows that $\gamma\partial_A\gamma^{-1}$ is a $\mathbf{K}$-derivation of R. It also follows from Lemma 3.13 that if $\mathcal{A}$ is nonempty the chain complex $(R, \gamma\partial_A\gamma^{-1})$ is acyclic. We show next how to construct this complex and prove that it is acyclic in the context of differential forms. To do this we use a close relative of the Koszul complex. The precise connection is described below.

Recall from Definition 7.1 that $\Omega(V)$ is the exterior algebra of the F-vector space $F \otimes V^*$. Thus $\Omega(V) = F \otimes \Lambda V^*$ where ΛV^* is the exterior algebra of the $\mathbf{K}$-vector space V^*. For $0 \leq p \leq \ell$ we have $\Omega^p(V) = F \otimes \Lambda^p V^*$. Let $x_1, \ldots, x_\ell$ be a basis for V^*.

DEFINITION 7.15. *Define an F-linear map*

$$\partial = \partial_\Omega : F \otimes \Lambda^p V^* \rightarrow F \otimes \Lambda^{p-1} V^*$$

by $\partial(f \otimes 1) = 0$ and for $1 \leq p \leq \ell$

$$\partial(f \otimes x_{i_1} \wedge \cdots \wedge x_{i_p}) = \sum_{k=1}^{p} (-1)^{k-1} x_{i_k} f \otimes (x_{i_1} \wedge \cdots \widehat{x_{i_k}} \cdots \wedge x_{i_p}).$$

The map ∂ is uniquely determined and satisfies $\partial^2 = 0$. Thus $(\Omega(V), \partial)$ is a chain complex. Note that $S \otimes V^*$ is stable under ∂ and hence $(S \otimes V^*, \partial)$ is a subcomplex. This subcomplex differs from the Koszul complex only in that when $p = 0$ here we have $f \otimes 1 \to 0$, while in the Koszul complex we have $f \otimes 1 \to \varepsilon(f)$ where $\varepsilon : S \to \mathbf{K}$ is the natural augmentation.

Lemma 7.16. *The map ∂ is a $\mathbf{K}$-derivation of $\Omega(V)$. For $\omega \in \Omega^p(V)$ and $\eta \in \Omega(V)$ we have:*

$$\partial(\omega\eta) = (\partial\omega)\eta + (-1)^p \omega(\partial\eta).$$

Proof. If $p = 0$ this is true because $\partial(f \otimes 1) = 0$. For $p \geq 1$ the fact that ∂ is an F-linear map allows us to consider only the cases $\omega = dx_{i_1} \cdots dx_{i_p}$ and $\eta = 1$ or $dx_{j_1} \cdots dx_{j_q}$, where the result follows by direct computation.

Proposition 7.17. *Let $\mathcal{A}$ be an arrangement and let $\partial = \partial_\Omega$. Then $\partial R(\mathcal{A}) \subset R(\mathcal{A})$. Let ∂_R denote the restriction of ∂ to $R(\mathcal{A})$. If $\mathcal{A}$ is nonempty then $(R(\mathcal{A}), \partial_R)$ is an acyclic chain complex and $\partial_R = \gamma\partial_A\gamma^{-1}$.*

Proof. It follows from Lemma 7.16 that for $(H_1, \ldots, H_p) \in \mathbf{S}$ we have

$$\partial(\omega_{H_1} \cdots \omega_{H_p}) = \sum_{k=1}^{p} (-1)^{k-1} \omega_{H_1} \cdots \widehat{\omega_{H_k}} \cdots \omega_{H_p}.$$

Thus $\partial R(\mathcal{A}) \subset R(\mathcal{A})$ so $(R(\mathcal{A}), \partial_R)$ is a chain complex. If $\mathcal{A}$ is nonempty, choose $H \in \mathcal{A}$. Let $\alpha = \alpha_H$ and let $\omega = \omega_H$. Write $\alpha = \sum c_k x_k$ for suitable $c_k \in \mathbf{K}$. It follows from Definition 7.15 that $\partial(dx_k) = \partial(1 \otimes x_k) = x_k \otimes 1$. Since ∂ is F-linear we have

$$\begin{aligned}
\partial\omega &= (1/\alpha)\partial(d\alpha) \\
&= (1/\alpha)\partial\left(\sum c_k dx_k\right) \\
&= (1/\alpha)\left(\sum c_k x_k\right) \otimes 1 \\
&= 1 \otimes 1 \\
&= 1
\end{aligned}$$

where in the last equality we identified $F \otimes \mathbf{K}$ with F. Now let $\eta \in R(\mathcal{A})$ be arbitrary. From Lemma 7.16 we get $\partial(\omega\eta) = (\partial\omega)\eta - \omega(\partial\eta) = \eta - \omega(\partial\eta)$. Suppose $\eta \in \ker \partial_R$. Since $R(\mathcal{A})$ is an algebra, $\omega\eta \in R(\mathcal{A})$ and thus $\eta = \partial_R(\omega\eta)$. This proves that $(R(\mathcal{A}), \partial_R)$ is acyclic. Finally note that $\gamma\partial_A\gamma^{-1}\omega_H = \gamma\partial_A a_H = \gamma 1 = 1 = \partial_R\omega_H$ for all $H \in \mathcal{A}$. Since both $\gamma\partial_A\gamma^{-1}$ and ∂_R are $\mathbf{K}$-derivations of $R(\mathcal{A})$ which agree on the generators ω_H of $R(\mathcal{A})$, they are equal.

Definition 7.18. *Let $\mathcal{A}$ be an arrangement with lattice $L = L(\mathcal{A})$. For $X \in L$ let $R_X = R_X(\mathcal{A}) = \sum \mathbf{K}\omega_{H_1} \cdots \omega_{H_p}$ where the sum is over all $(H_1, \ldots, H_p) \in \mathbf{S}_X$. Let $R_p = \sum_{X \in L_p} R_X$.*

PROPOSITION 7.19. *We have*

$$R_p = \bigoplus_{X \in L_p} R_X.$$

Proof. The sum is direct because $R_X = \gamma(A_X)$, γ is an isomorphism, and $A_p = \bigoplus_{X \in L_X} A_X$ by Corollary 3.18.

8 The Topology of $M(\mathcal{A})$

In this section we discuss recent developments in the study of the topology of the complement, not covered in the survey article by Falk and Randell [46].

Salvetti's Complex

Let $(\mathcal{B}, W)$ be a real ℓ-arrangement, which may be affine, and let $(\mathcal{A}, V)$ be its complexification. In [126] Salvetti constructed a finite CW-complex $X \subset M(\mathcal{A})$ such that the inclusion is a homotopy equivalence. Here we shall outline the construction without proofs, and with different notation.

The hyperplanes of $\mathcal{B}$ subdivide $W = \mathbf{R}^\ell$ into **facets**. Codimension 0 facets are chambers defined in the Introduction. Codimension 1 facets are called **faces**. (Some authors use the terms "face" and "facet" with interchanged meaning.) Let $\mathcal{F}$ denote the set of all facets. The support $|P|$ of a facet P is the affine subspace generated by P. Each facet is open in its support. Let $\overline{P}$ denote the closure of P. We say that the facet P **contains** the facet Q if $\overline{P} \supset Q$. A hyperplane $H \in \mathcal{B}$ is a **wall** of the chamber C if it contains one of the faces of C. Call two chambers **adjacent** if they have a common face.

LEMMA 8.1. (i) *Suppose H is a wall of C. Then there is a unique chamber C' which is adjacent to C and has H as a wall. Moreover, H is the only hyperplane which separates C and C'.*

(ii) *Let C_1, C_2, C_3 be three chambers and let $\mathcal{B}(C_i, C_j)$ be the set of hyperplanes which separate C_i and C_j. Then*

$$\mathcal{B}(C_1, C_2) = [\mathcal{B}(C_1, C_3) - \mathcal{B}(C_2, C_3)] \cup [\mathcal{B}(C_2, C_3) - \mathcal{B}(C_1, C_3)].$$

Given $Q \in \mathcal{F}$ let

$$\mathrm{pr}_Q : W \to W/|Q|$$

be the affine projection. It sends the affine hyperplanes which contain Q into hyperplanes of a central arrangement $(\mathcal{B}/Q, W/|Q|)$. Let $\mathcal{F}/Q$ be the set of facets of $\mathcal{B}/Q$. Let

$$\pi_Q : \mathcal{F} \to \mathcal{F}/Q$$

be the map which sends $P \in \mathcal{F}$ to the smallest facet of $\mathcal{F}/Q$ which contains $\mathrm{pr}_Q(P)$.

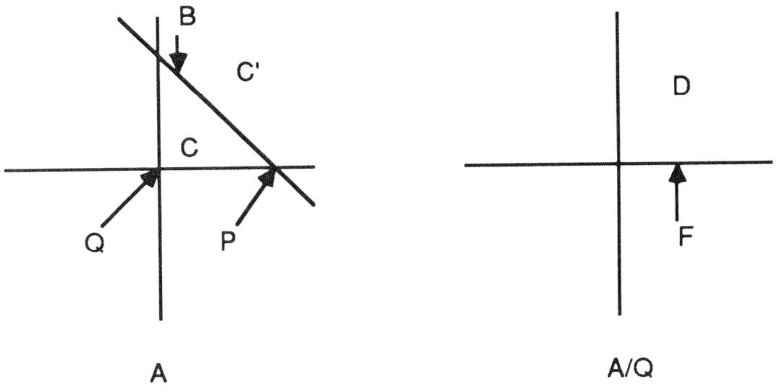

Figure 20: Example of π_Q

If $P \in \mathcal{F}$ then $\operatorname{codim}_W(P) \geq \operatorname{codim}_{W/|Q|}(\pi_Q(P))$. Thus π_Q sends chambers to chambers. Let $\mathcal{F}_Q \subset \mathcal{F}$ be the set of facets which contain Q. Then the restriction $\pi_Q : \mathcal{F}_Q \to \mathcal{F}/Q$ is a bijection which preserves codimension. An illustration of the map π_Q is given in Figure 20 where Q is the origin. The chambers C , C' and the face B map to the chamber D. The facet P maps to the face F.

Define a map $w : \mathcal{F} \to W$ which sends each facet Q to a point $w(Q) \in Q$. Let $\mathcal{V}^i$ be the set of images of codimension i facets, and let $\mathcal{V} = \bigcup \mathcal{V}^i$. Then the restriction $w : \mathcal{F} \to \mathcal{V}$ is a bijection.

Let G be the undirected graph which has one vertex for each chamber, and an edge between two vertices if the corresponding chambers are adjacent. Let VG be the vertex set and let EG be the edge set for G. In order to embed G in W we need the following notation. Given two points $u_0, u_1 \in W$, their **join** $u_0 * u_1$ is the line segment between u_0 and u_1:

$$u_0 * u_1 = \{(1-\lambda)u_0 + \lambda u_1,\ \lambda \in [0,1]\}.$$

This may be iterated for the affine independent points $u_0, \ldots, u_k$, $k \leq \ell$ to obtain their convex hull, a k-simplex denoted $u_0 * \cdots * u_k$. Identify VG with $\mathcal{V}^0$. If C and C' are adjacent chambers with common face P then the corresponding edge is $w(C) * w(P) \cup w(P) * w(C')$. Let CG be the set of edge-paths of G. Given a path $c \in CG$ define its **length** $\ell(c)$ to be the number of edges in c. Call c **minimal** if its length is minimum among the paths joining the ends of c.

LEMMA 8.2. *A path c between $w(C)$ and $w(C')$ is minimal if and only if it crosses once and only once the hyperplanes which separate C and C', and none of the other hyperplanes. Thus $\ell(c) = |\mathcal{B}(C, C')|$.*

For $Q \in \mathcal{F}$ let $VG_Q \subset \mathcal{F}_Q$ be the set of vertices for chambers which contain Q, and let EG_Q be the set of edges which cross faces of $\mathcal{F}_Q$.

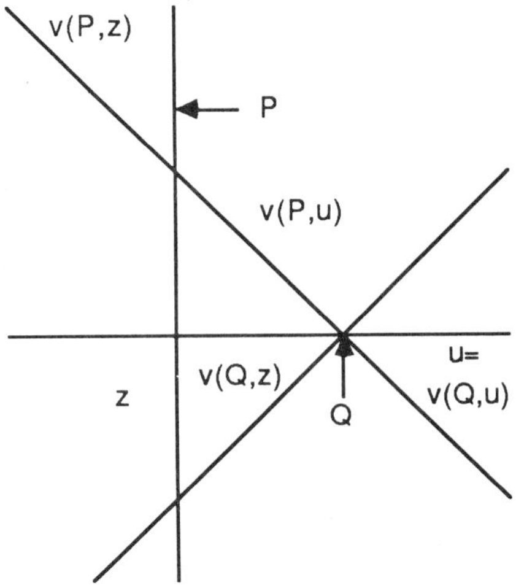

Figure 21: Examples of the map v

LEMMA 8.3. (i) *Given $Q \in \mathcal{F}$ and $z \in VG$, there exists a unique vertex $v(Q,z) \in VG_Q$ which has the shortest distance from z.*

(ii) *If $z \in VG$ and $u \in VG_Q$ then a minimal path in CG between z and u is the composition of a minimal path c between z and $v(Q,z)$ with a minimal path d between $v(Q,z)$ and u.*

(iii) *The path c does not cross any hyperplane which contains Q, and the edges of d are in EG_Q.*

The map $v : \mathcal{F} \times VG \to VG$ where $(Q,z) \to v(Q,z)$ gives rise to two maps:

$$\begin{array}{lll} v_Q : VG \to VG_Q & \text{where} & z \to v(Q,z), \\ v_z : \mathcal{F} \to VG & \text{where} & Q \to v(Q,z). \end{array}$$

Introduce a partial order on the facets by *inclusion*. Thus $P \geq Q$ if P contains Q. Since $w : \mathcal{F} \to \mathcal{V}$ is a bijection, this induces a partial order on $\mathcal{V}$ by $w(P) \geq w(Q) \Leftrightarrow P \geq Q$. Write Q^k when $Q \in \mathcal{F}$ has codimension k. Note that the partial order defined in Section 3 differs from this one in two aspects. It was defined on the set $L(\mathcal{B})$, which is the set of supports, not the set of facets; and it was by reverse inclusion.

LEMMA 8.4. *Suppose $\mathcal{B}$ is an essential central arrangement with $T(\mathcal{B}) = \{0\}$. Let $Q \neq \{0\}$ be a facet and let $x \in Q$. For every $y \in Q$ there exists a number $t_0 > 0$ such that the half-line from x through ty, $t \geq 0$, meets $\partial\overline{Q}$ if $t < t_0$, and does not meet any hyperplane of $\mathcal{B} - \mathcal{B}_Q$ if $t \geq t_0$.*

Thus in a central arrangement $\mathcal{B}$ with $T(\mathcal{B}) = \{0\}$, the closure of each facet is a cone over a polyhedron with vertex the origin. The next lemma gives a precise description of the base polyhedron.

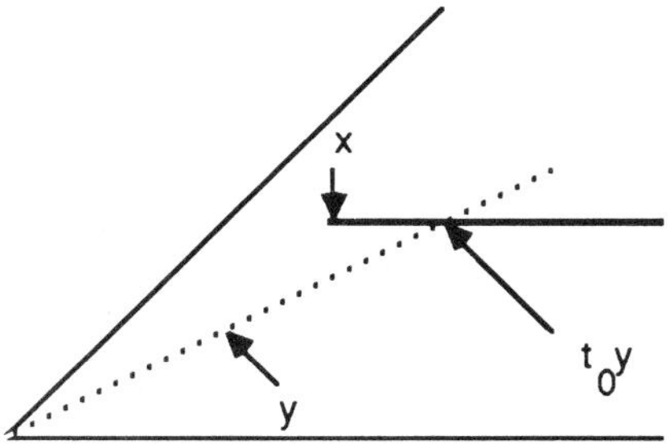

Figure 22: The critical half-line

LEMMA 8.5. *Suppose* $\mathcal{B}$ *is an essential central arrangement with* $T(\mathcal{B}) = \{0\}$. *Given a point* $x \in W - \{0\}$

(i) *there exists a unique chain of facets* $Q^{j_0} > Q^{j_1} > \cdots > Q^{j_k}$ *such that the half-line* $s(x) = \{tx | t > 0\}$ *intersects the simplex* $w(Q^{j_0}) * \cdots * w(Q^{j_k})$, *and*

(ii) *the intersection is a single point* x' *which may be expressed in barycentric coordinates as* $x' = \sum \lambda_i w(Q^{j_i})$ *where* $\sum \lambda_i = 1$, *and* $\lambda_i \in (0, 1]$ *for* $0 \le i \le k$.

In a central ℓ-arrangement these simplexes fit together to form an $(\ell - 1)$-sphere. To see this consider the map $x' \to x'/|x'|$. Assign to the origin the chain consisting of the facet $\{0\}$. The next result shows that in an arbitrary arrangement the simplexes corresponding to a fixed facet fit together in a similar way.

LEMMA 8.6. *Let* $P^k \in \mathcal{F}$ *and let* S *be the set of all chains* $Q^0 > \cdots > Q^{k-1} > P^k$. *Then*

$$D_P^k = \bigcup_S w(Q^0) * \cdots * w(Q^{k-1}) * w(P^k)$$

is a triangulated k*-cell in* V *whose boundary is*

$$S_P^{k-1} = \bigcup_S w(Q^0) * \cdots * w(Q^{k-1}).$$

Note that we assigned to the facet P^k of *codimension* k a cell D_P^k of *dimension* k. This seemingly confusing notation is in fact quite natural. Let

$$A = \bigcup_{P^k \in \mathcal{F}} D_P^k.$$

Then A is a cellular complex with a barycentric triangulation. We may view P^k as an $(\ell - k)$-cell of the cell complex $\mathcal{F}$ whose union is W. Its dual cell complex is A and D_P^k is the cell dual to P^k. Figure 23 indicates some of these dual cells. The next lemma shows that all facets of minimal dimension are parallel subspaces.

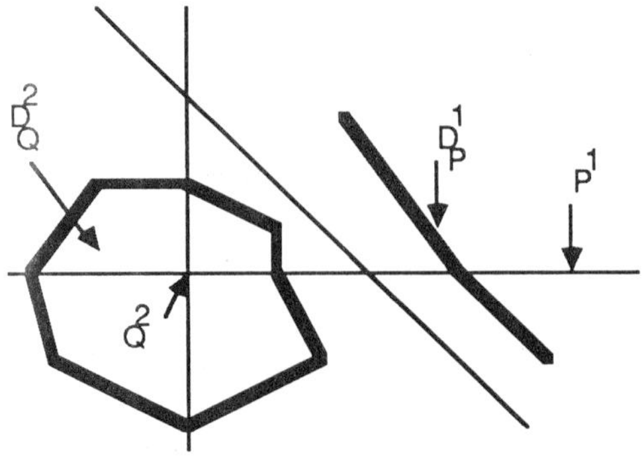

Figure 23: Dual cells

Lemma 8.7. *Let p be the maximum codimension of facets in $\mathcal{F}$ and let $Q^p \in \mathcal{F}$. Then for every $H \in \mathcal{B}$ either $Q^p \subset H$ or H is parallel to Q^p. Thus $|Q^p| = Q^p$. Conversely, if Q is a facet and for every $H \in \mathcal{B}$ either $Q \subset H$ or H is parallel to Q, then Q has codimension p and Q is parallel to Q^p.*

Lemma 8.8. *Let p be the maximum codimension of facets in $\mathcal{F}$ and let $P^j \in \mathcal{F}$. There exist two chains $Q^0 > \cdots > Q^{j-1}$ and $Q^{j+1} > \cdots > Q^p$, which are in general not unique, such that*

$$Q^0 > \cdots > Q^{j-1} > P^j > Q^{j+1} > \cdots > Q^p.$$

If $p = \ell$ then the arrangement is essential and the facets of maximal codimension are points. Let B be the union of bounded facets. Then A is a neighborhood of B and B is a deformation retract of A. In fact B is a deformation retract of W. It follows from Lemma 8.8 that $\overline{P} \cap B \neq \emptyset$ for every facet P. It follows from Lemma 8.5 that for an unbounded facet $\overline{P} = \overline{P} \cap B \times \mathbf{R}^+$. Thus B is contractible and hence A is contractible.

If $p < \ell$, let Q^p be a facet of maximal codimension. Write $W = |Q^p| \oplus S$ with some subspace S. The hyperplanes of $\mathcal{B}$ determine an arrangement in S. It follows from Lemma 8.7 that $P^k \cap S$ has codimension k in S. Call $P \in \mathcal{F}$ relatively bounded if $P \cap S$ is bounded. Let $B^{(r)}$ be the union of the relatively bounded facets in W, and let B_S be the union of the bounded facets in S. There is a deformation retraction of the pair $(W, B^{(r)})$ onto (S, B_S). Thus $B^{(r)}$ is contractible, and hence A is contractible.

Recall that $(V, \mathcal{A})$ is the complexification of $(W, \mathcal{B})$. Write $z \in V$ as $z = x + iy$ with $x, y \in W$ and let $\Re$, $\Im$ denote real and imaginary parts. For each hyperplane $H \in \mathcal{B}$ let H_0 be the hyperplane in W parallel to H through the origin. Thus H_0 is a vector subspace. Let $z = x + iy \in V$. Then $z \in M(\mathcal{A})$ if and only if for every $H \in \mathcal{B}$ with $x \in H$, $y \notin H_0$.

Salvetti's CW-complex $X \subset M(\mathcal{A})$ is constructed as follows. Let $P^k \in \mathcal{F}$. Identify the standard k-cell D^k with D^k_P of Lemma 8.6. Write $x \in D^k_P$ in barycentric coordinates as $x = \sum_{i=0}^{k} \lambda_i w(P^i)$, $\sum \lambda_i = 1$, $\lambda_i \in [0,1]$, where $P^0 > \cdots > P^k$. For each $u \in VG$ define an embedding $\Phi_{P,u} : D_P^k \to M(\mathcal{A})$ by

$$\Re\Phi_{P,u}(x) = x,$$

$$\Im\Phi_{P,u}(x) = \sum_{i=0}^{k} \lambda_i [v(P^i, u) - w(P^i)].$$

The definition of the real part shows that this is an embedding into V. The image is clearly in $M(\mathcal{A})$ if $x \in M(\mathcal{B})$. If $x \notin M(\mathcal{B})$ then $x \in H$ for some $H \in \mathcal{B}$ and $w(P^0)$ appears in the barycentric coordinates of x such that H is a wall of P^0. Thus $\lambda_0 = 0$ but some $\lambda_i > 0$. Clearly when $\lambda_i > 0$ we have $P^i \subset H$. It follows from Lemma 8.3.iii applied to P^i that $v(P^i, u)$ and u are on the same side of H. Thus the points $v(P^i, u) - w(P^i)$ are on the same side of H_0. Since $y = \Im\Phi_{P,u}(x)$ is a positive linear combination of these points, $y \notin H_0$. Thus $\Phi_{P,u}(D_P^k) \subset M(\mathcal{A})$.

Let $D_{P,u}^k = \Phi_{P,u}(D_P^k)$ be this embedded k-cell. If $u, u' \in VG$ are such that $v(P,u) = v(P,u')$ then $D_{P,u}^k = D_{P,u'}^k$. Since for a fixed facet P the image of the map v_P is in VG_P, the number of distinct cells attached by $\Phi_{P,u}$ is $|VG_P|$. Let

$$X^k = \bigcup D_{P,u}^k$$

where the union is over all $P^k \in \mathcal{F}$ and $u \in VG$, and let

$$X = \bigcup_{k \geq 0} X^k.$$

Then X is a regular cell complex with k-skeleton X^k. Note that for a chamber $P = P^0$ we have $D_{P,u}^0 = w(P) + i(v(P,u) - w(P))$. Since $v(P,u) = w(P)$ we get $D_{P,u}^0 = w(P)$. Thus each chamber P gives rise to one 0-cell, $w(P)$.

Figure 24 illustrates this complex for the 1-arrangement whose only hyperplane is the origin. There are two chambers C, C' and one face P. Let $w(C) = +1$, $w(C') = -1$, and $w(P) = P = 0$. It follows that $X^0 = \{w(C), w(C')\} = \{+1, -1\}$. The dual cell of the face P^1 is $D_P^1 = w(C) * w(P) \cup w(P) * w(C')$. Since $VG_P = \{w(C), w(C')\}$, P contributes two 1-cells to X. Consider the cell $D_{P,w(C)}^1$. Suppose $x \in w(C) * w(P)$. Then $x = \lambda_0 w(C) + \lambda_1 w(P)$, where $\lambda_0 + \lambda_1 = 1$, and $\lambda_i \in [0,1]$. We have

$$\Re\Phi_{P,w(C)}(x) = x,$$

$$\Im\Phi_{P,w(C)}(x) = \lambda_0[v(C, w(C)) - w(C)] + \lambda_1[v(P, w(C)) - w(P)].$$

Note that $v(C, w(C)) = w(C) = v(P, w(C))$. Thus if we set $\lambda_0 = t$ and use $w(P) = 0$ and $w(C) = 1$ we get

$$\Phi_{P,w(C)}(t) = t + i(1-t) \quad \text{for } t \in [0,1].$$

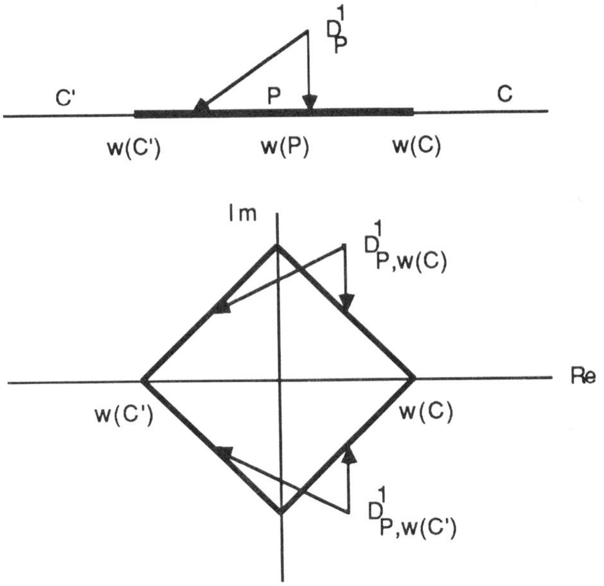

Figure 24: Example of the complex X

Similarly, if $x \in w(C') * w(P)$ write $x = \lambda_0 w(C') + \lambda_1 w(P)$. Then

$$\Re\Phi_{P,w(C)}(x) = x,$$

$$\Im\Phi_{P,w(C)}(x) = \lambda_0[v(C', w(C)) - w(C')] + \lambda_1[v(P, w(C)) - w(P)].$$

Here $v(C', w(C)) = w(C')$ and $v(P, w(C)) = w(C)$. Thus if we set $\lambda_0 = t$ and use $w(C') = -1$, $w(P) = 0$, $w(C) = 1$ we get

$$\Phi_{P,w(C)}(-t) = -t + i(1-t) \quad \text{for } t \in [0, 1].$$

The cell $D^1_{P,w(C')}$ is computed similarly. The main result of [126, Part I] is:

THEOREM 8.9. *The inclusion $X \to M(\mathcal{A})$ is a homotopy equivalence.*

In principle the fundamental group may be computed for any arrangement. The difficulty is a matter of bookkeeping. Randell [116] gave a presentation for the fundamental group of the complement of a complexified real arrangement. Salvetti [126, Part II] gave a different presentation of $\pi_1(M(\mathcal{A}))$ with one generator for each hyperplane, and one set of relations for each codimension 2 subspace. At the time of this writing there is no algorithm for the presentation of $\pi_1(M(\mathcal{A}))$ for an arbitrary complex arrangement, other than the classical method using a pencil of lines. In a recent preprint Naruki [96] gave a presentation of the fundamental group of the complement of the unitary reflection arrangement which arises from Klein's simple group of order 168. The $K(\pi, 1)$ problem is still open for this space.

Minimal Models

Next we describe work of Kohno [77], [78] and Falk [42] who use the theory of minimal models to establish a connection between $\pi_1(M(\mathcal{A}))$ and $H^*(M(\mathcal{A}))$. First we define the terminology and describe some general results of Sullivan [138] and Morgan [95]. Griffiths and Morgan gave a detailed exposition of this work in [59].

All vector spaces and algebras are over $\mathbf{Q}$. A differential graded algebra (DG algebra) A is a graded vector space $A = \bigoplus_{i\geq 0} A^i$ with a degree 1 coboundary operator $d : A^i \to A^{i+1}$, and a product $\wedge : A^i \otimes A^j \to A^{i+j}$. These satisfy:

(i) $d^2 = 0$,

(ii) $d(x \wedge y) = dx \wedge y + (-1)^i x \wedge dy$ for $x \in A^i$,

(iii) $x \wedge y = (-1)^{ij} y \wedge x$ for $x \in A^i$ and $y \in A^j$,

(iv) $\wedge$ makes A an associative algebra with unit, $1 \in A^0$.

If $A^0 = \mathbf{Q}$ then A is called connected. The augmentation ideal of a connected algebra is $\mathcal{I}(A) = \bigoplus_{i>0} A^i$. The quotient $I(A) = \mathcal{I}(A)/\mathcal{I}(A) \wedge \mathcal{I}(A)$ is the set of indecomposable elements.

Let V be a finite dimensional vector space. Let $\Lambda_r(V)$ be the graded-commutative algebra over V where $\deg v = r$ for $v \in V$. Thus $\Lambda_r(V)$ is a polynomial algebra if r is even, and an exterior algebra if r is odd. Note that in this case $I(A) = V$.

Let B be a DG algebra. Let A be a DG subalgebra of B. The inclusion $A \subset B$ is called a **Hirsch extension** of degree r if for some V there is an isomorphism of graded-commutative algebras $B \simeq A \otimes \Lambda_r(V)$, and the differential d of B satisfies $dV \subset A^{r+1}$.

Let (M, d) be a DG algebra. It is called **minimal** if

(i) $M^0 = \mathbf{Q}$,

(ii) there is an increasing filtration of DG subalgebras

$$\mathbf{Q} = M_0 \subset M_1 \subset M_2 \subset \cdots$$

such that $M = \bigcup M_i$, and each inclusion $M_i \subset M_{i+1}$ is a Hirsch extension,

(iii) d is decomposable: $dM \subset \mathcal{I}(M) \wedge \mathcal{I}(M)$.

Let A be a DG algebra. An i-**minimal** model for A is a map $\rho : M \to A$ of DG algebras such that:

(i) M is minimal,

(ii) M is generated by elements of degree $\leq i$,

(iii) $\rho^* : H^p(M) \to H^p(A)$ is an isomorphism for $p \leq i$ and injective for $p = i + 1$.

In case $i = \infty$, $\rho : M \to A$ is called a **minimal model** for A. It follows from the work of Sullivan [138] and Morgan [95] that if A is a connected DG algebra then an i-minimal model for A exists for each i and it is unique up to isomorphism.

Suppose X is a connected polyhedron. Sullivan defined the DG algebra $A = A(X)$ of $\mathbf{Q}$-polynomial forms on X. The 1-minimal model $\mathcal{M} \to A$ is an increasing union of Hirsch extensions of degree 1

$$\mathbf{Q} = \mathcal{M}_0 \subset \mathcal{M}_1 \subset \mathcal{M}_2 \subset \cdots,$$

$$\mathcal{M} = \bigcup_{n \geq 0} \mathcal{M}_n.$$

Let V_n be the degree 1 part of $\mathcal{M}_n$. The differential in $\mathcal{M}_n$ is determined by its restriction to V_n. Since $\mathcal{M}$ is minimal, we have:

$$d|V_n : V_n \to V_n \wedge V_n.$$

The Lie algebra $\mathcal{L}_n$ dual to $\mathcal{M}_n$ has underlying vector space V_n^*. The bracket is dual to $d|V_n$

$$[\ ,\ ] : V_n^* \wedge V_n^* \to V_n^*.$$

The inclusions $\mathcal{M}_n \subset \mathcal{M}_{n+1}$ give rise to maps $\mathcal{L}_n \leftarrow \mathcal{L}_{n+1}$. Induction shows that the $\mathcal{L}_n$ are nilpotent. This constructs from the filtration of the 1-minimal model $\mathcal{M}$ a tower of nilpotent Lie algebras:

$$0 \leftarrow \mathcal{L}_1 \leftarrow \mathcal{L}_2 \leftarrow \cdots.$$

Let G be a finitely presented group. The **lower central series** G_n of G is defined by setting $G_0 = G$ and $G_n = [G_{n-1}, G]$ for $n \geq 1$. Here $[G_k, G]$ denotes the subgroup generated by elements of the form $xyx^{-1}y^{-1}$ with $x \in G_k$, $y \in G$. The quotients G_{n-1}/G_n are finitely generated abelian groups. Let $\phi_n(G) = \operatorname{rank}(G_{n-1}/G_n)$. The quotients G/G_n are nilpotent groups. By a construction of Malcev, see [59, pp.142-145] it is possible to "tensor" these nilpotent groups by $\mathbf{Q}$ and use the central extensions

$$0 \to G_{n-1}/G_n \to G/G_n \to G/G_{n-1} \to 1$$

to define a Lie algebra structure on $(G/G_n) \otimes \mathbf{Q}$ called $\mathcal{L}_n(G)$. If $G = \pi_1(X)$ this leads to Sullivan's theorem [138], [95, (5.11)], [59, p.145].

THEOREM 8.10. *Let X be a connected polyhedron, let $\rho : \mathcal{M} \to A(X)$ be a 1-minimal model and let*

$$0 \leftarrow \mathcal{L}_1 \leftarrow \mathcal{L}_2 \leftarrow \cdots$$

be the tower of dual nilpotent Lie algebras. Let $G = \pi_1(X)$. Then $\mathcal{L}_n \simeq \mathcal{L}_n(G)$ for $n \geq 0$.

Since $\mathcal{M}_n$ is a degree 1 Hirsch extension of $\mathcal{M}_{n-1}$ we have $\mathcal{M}_n \simeq \mathcal{M}_{n-1} \otimes \Lambda_1(W_n)$ for some vector space W_n. The following is a direct consequence of Theorem 8.10.

COROLLARY 8.11. *Let X be a connected polyhedron with finitely generated rational cohomology. Let $\rho : \mathcal{M} \to A(X)$ be a 1-minimal model and let $G = \pi_1(X)$. Then $\phi_n(G) = \dim W_n$.*

Now suppose $\mathcal{A}$ is an arrangement and $M = M(\mathcal{A})$. Since M is a formal space in the sense of Sullivan [138, p.315], we may replace the algebra $A(M)$ of **Q**-polynomial forms on M with the algebra $A = A(\mathcal{A})$. Let $\rho : \mathcal{M} \to A$ be a 1-minimal model. We may view A as a DG algebra with zero differential so $H^*(A) = A$.

DEFINITION 8.12. *Call $\mathcal{A}$ a* **rational** *$K(\pi, 1)$-arrangement if $\rho^* : H^*(\mathcal{M}) \to A$ is an isomorphism.*

Falk [42] and Kohno [81] used different methods to prove the following:

THEOREM 8.13. *Let $\mathcal{A}$ be a rational $K(\pi, 1)$-arrangement and write $\phi_n = \phi_n(\pi_1(M))$. Then*

$$\prod_{n \geq 1} (1 - t^n)^{\phi_n} = P(M, -t).$$

This formula is called the LCS (lower central series) formula. It connects the ranks of the successive quotients in the lower central series of the fundamental group of M with the Poincaré polynomial of M. It is natural to ask for the largest class of arrangements for which the LCS formula is valid. Falk and Randell [45] showed that it holds for fiber type arrangements. It is also known to hold for certain reflection arrangements, and to be false for some reflection arrangements.

Discriminantal Arrangements

In a recent preprint Manin and Schechtman [90] constructed a family of arrangements which generalize the braid arrangements. Let $W = \mathbf{K}^k$ and let $\mathcal{A}^0 = \{H_1^0, \ldots, H_n^0\}$ with $n > k$ be an affine general position arrangement in W. Let $U(n, k)$ be the set of k-arrangements $\mathcal{A} = \{H_1, \ldots, H_n\}$ in W which satisfy:

(i) H_i is parallel to H_i^0 for $1 \leq i \leq n$,

(ii) $\mathcal{A}$ is a general position arrangement.

Manin and Schechtman showed that $U(n, k)$ is itself the complement of an arrangement. Let $V = \mathbf{K}^n$ be the space of parallel translations of the hyperplanes of $\mathcal{A}^0$. Denote by $C(n, a)$ the set of subsets of $\{1, \ldots, n\}$ of cardinality a. For $K \in C(n, a)$ let D_K be the set of parallel translations $(H_1, \ldots, H_n) \in V$ such that $\bigcap_{i \in K} H_i \neq \emptyset$. If $|K| \leq k$ then $D_K = V$ and if $|K| \geq k + 1$ then $\operatorname{codim} D_K = |K| - k$. In particular for $J \in C(n, k + 1)$ the set D_J is a hyperplane in V, and these hyperplanes are pairwise distinct. Let $\mathcal{B}(n, k) = \{D_J | J \in C(n, k + 1)\}$. The arrangement $\mathcal{B}(n, 1)$ is the braid arrangement.

Manin and Schechtman called $\mathcal{B}(n,k)$ a **discriminantal arrangement**. In our terminology they proved that

$$U(n,k) = M(\mathcal{B}(n,k)).$$

Although $\mathcal{B}(n,k)$ also depends on $\mathcal{A}^0$, its combinatorial properties do not. Let $L(n,k) = L(\mathcal{B}(n,k))$ and let $\ell = n-k$. It is easy to see that $L(n,k)$ has a unique maximal element $D_{(1,\dots,n)}$ of codimension ℓ and hence $r(B(k+\ell,k)) = \ell$. In fact we may identify $D_{(1,\dots,n)}$ with W. Assume that $H_1^0,\dots,H_k^0$ contain the origin of W. For every $w \in W$ there is a parallel translation such that every hyperplane contains the endpoint of w. Thus for $\ell = 1$ and $\ell = 2$ the lattices $L(k+1,k)$ and $L(k+2,k)$ are easy to describe.

Manin and Schechtman gave the following description of $L(k+3,k)$. For $J = (1,\dots,k+3) - (i,j)$ write $D_J = (i,j)$. For $K = (1,\dots,k+3) - (i)$ write $D_K = (i)$. If i,j,l,m are distinct indices write $(ij,lm) = (i,j) \cap (l,m)$. The hyperplanes of $\mathcal{B}(k+3,k)$ are the $(k+2)(k+3)/2$ sets (i,j). The elements of $L(k+3,k)$ of rank 2 are the $k+3$ sets (i) and the $k(k+1)(k+2)(k+3)/8$ sets (ij,lm). Each (i) is contained in $k+2$ hyperplanes, and each (ij,lm) is contained in 2 hyperplanes. Let T denote the unique maximal element of rank 3. This gives:

$$\begin{aligned}
\mu(V) &= 1,\\
\mu((i,j)) &= -1,\\
\mu((i)) &= k+1,\\
\mu((ij,lm)) &= 1,\\
\mu(T) &= -(1/8)k(k+3)(k^2+3k+6) - 1.
\end{aligned}$$

PROPOSITION 8.14. *The Poincaré polynomial of* $\mathcal{B}(k+3,k)$ *is:*

$$\begin{aligned}
\pi(\mathcal{B}(k+3,k),t) = 1 + (1/2)(k+2)(k+3)t + (1/8)(k+1)(k+3)\\
\cdot(k^2+2k+8)t^2 + ((1/8)k(k+3)(k^2+3k+6)+1)t^3.
\end{aligned}$$

It is not known whether the arrangements $\mathcal{B}(n,k)$ are $K(\pi,1)$. The first nontrivial cases are the arrangements $\mathcal{B}(k+3,k)$ for $k \geq 2$. It is immediate from the description above that they may be visualized as follows.

Let $\mathcal{C}^*(n)$ be an affine real 2-arrangement consisting of n points in the plane labeled (i) for $1 \leq i \leq n$ with the following properties:

(i) no three points are collinear,

(ii) if $(i),(j),(l),(m)$ are distinct points then the line through $(i),(j)$ is not parallel to the line through $(l),(m)$.

Let $\mathcal{C}(n)$ be the central 3-arrangement obtained by embedding $\mathcal{C}^*(n)$ as the affine set $z=1$ and coning over the origin. Thus if Q^* defines $\mathcal{C}^*(n)$ then its homogenization Q defines $\mathcal{C}(n)$. It follows from the description that

$$\mathcal{B}(k+3,k) \approx \mathcal{C}(k+3) \times \Phi_k$$

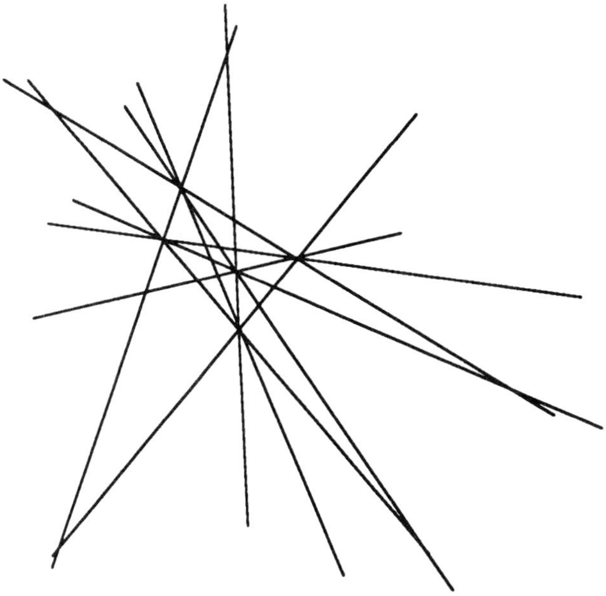

Figure 25: The arrangement $\mathcal{C}^*(5)$

where Φ_k is the empty k-arrangement. The arrangement $\mathcal{C}^*(5)$ is illustrated in Figure 25. Using Terao's factorization theorem we show in Proposition 9.20 that the arrangements $\mathcal{B}(k+3,k)$ are not free for $k \geq 2$.

9 Free Arrangements

In this section we describe some of Terao's work on free arrangements. The proofs require more commutatve algebra than we could present in this section. A complete account will appear in [111]. We return to the convention that all arrangements are central.

The Module of $\mathcal{A}$-derivations

Recall that if $x_1, \ldots, x_\ell$ is a basis for V^*, then the polynomial algebra is $S = \mathbf{K}[x_1, \ldots, x_\ell]$. It has a natural grading by polynomial degree, $S = \bigoplus_p S_p$ where $S_0 = \mathbf{K}$ and $S_1 = V^*$. We defined the module of derivations, $\mathrm{Der}_{\mathbf{K}}(S)$ in the Introduction. It is a free S-module with basis $D_i = \partial/\partial x_i$ for $1 \leq i \leq \ell$. It is graded by saying that $\theta \in \mathrm{Der}_{\mathbf{K}}(S)$ has degree q if $\theta(S_p) \subset S_{p+q}$. With this grading $\deg D_i = -1$.

Definition 9.1. *Let $f \in S$ and define*

$$D_S(f) = \{\theta \in \mathrm{Der}_{\mathbf{K}}(S) | \theta(f) \in fS\}.$$

If $\mathcal{A}$ is an arrangement with defining polynomial $Q = Q(\mathcal{A})$ its module of $\mathcal{A}$-derivations is $D_S(\mathcal{A}) = D_S(Q)$.

Proposition 9.2. *If $f_1, f_2 \in S$ are coprime, then*

$$D(f_1 f_2) = D(f_1) \cap D(f_2).$$

Proof. Let $\theta \in Der_{\mathbf{K}}(S)$. Then

$$\begin{aligned} \theta \in D(f_1 f_2) &\Leftrightarrow \theta(f_1 f_2) \in f_1 f_2 S \\ &\Leftrightarrow f_1 \theta(f_2) + f_2 \theta(f_1) \in f_1 f_2 S \\ &\Leftrightarrow \theta(f_i) \in f_i S \quad \text{for } i = 1, 2 \\ &\Leftrightarrow \theta \in D(f_1) \cap D(f_2). \end{aligned}$$

Corollary 9.3. (i) *If $Q = \prod_{H \in \mathcal{A}} \alpha_H$ then*

$$D(\mathcal{A}) = D(Q) = \bigcap_{H \in \mathcal{A}} D(\alpha_H).$$

(ii) *If $\mathcal{B} \subset \mathcal{A}$ is a subarrangement then $D(\mathcal{B}) \supset D(\mathcal{A})$.*

We describe the topological significance of the module $D(\mathcal{A})$. Since $N = \bigcup_{H\in\mathcal{A}} H$ is a singular variety, it has no tangent space. But V has a stratification induced by N and each stratum has a tangent space. We show next that at every point of V the evaluation of $D(\mathcal{A})$ at the point spans the tangent space there.

For $X \in L(\mathcal{A})$ let $M(X) = M(\mathcal{A}^X)$. Then $M(X)$ is an open submanifold of X and we have a disjoint union:

$$V = \bigcup_{X\in L} M(X).$$

For $v \in V$ let TV_v denote the tangent space of V at v. Then TV_v is a $\mathbf{K}$ vector space with basis D_i. Thus we can define an evaluation map $\rho_v : D(\mathcal{A}) \to TV_v$ as follows. Given $\theta \in D(\mathcal{A})$ write $\theta = \sum h_i D_i$ and let $\rho_v(\theta) = \sum h_i(v) D_i$. Write $D(\mathcal{A})_v = \rho_v(D(\mathcal{A}))$.

PROPOSITION 9.4. *If $v \in M(X)$ then $D(\mathcal{A})_v = TM(X)_v$.*

Proof. Suppose $r(X) = p$. Choose coordinates so that $X = H_1 \cap \cdots \cap H_p$ where $H_j = \ker x_j$ for $1 \le j \le p$. Note that this makes $Q(\mathcal{A})$ divisible by $x_1 \cdots x_p$. We may choose D_i for $p+1 \le i \le \ell$ as a basis for $TX_v = TM(X)_v$. In the rest of this paragraph let $1 \le j \le p$. If $v \in M(X)$ then $x_j(v) = 0$. If $\theta \in D(\mathcal{A})$ then by Corollary 9.3 $\theta \in D(x_j)$ and hence $\theta(x_j) \in x_j S$. Write $\theta = \sum_{i=1}^{\ell} h_i D_i$. It follows that h_j is divisible by x_j and hence $h_j(v) = 0$. Thus $\rho_v(\theta) = \sum_{i=p+1}^{\ell} h_i(v) D_i \in TM(X)_v$.

For the converse let $Q = Q_1 Q_2$ where $Q_1 = Q(\mathcal{A}_X)$ and $Q_2 = Q(\mathcal{A} - \mathcal{A}_X)$. By our choice of coordinates above, $Q_1 \in \mathbf{K}[x_1, \ldots, x_p]$ and if $v \in M(X) \subset X$ then $Q_2(v) = c \neq 0$. It is easy to check that for $p+1 \le i \le \ell$ we have $\theta_i = Q_2 D_i \in D(\mathcal{A})$. Since $\rho_v \theta_i = c D_i$ it follows that $TM(X)_v \subset D(\mathcal{A})_v$.

DEFINITION 9.5. *The arrangement $\mathcal{A}$ is free if $D_S(\mathcal{A})$ is a free S-module.*

The empty arrangement Φ_ℓ is free with basis $\{D_1, \ldots, D_\ell\}$. This is clear since $D_S(\Phi_\ell) = \mathrm{Der}_{\mathbf{K}}(S)$. The Boolean arrangement is free with basis $\{x_1 D_1, \ldots, x_\ell D_\ell\}$. To see this recall that $Q = x_1 \cdots x_\ell$. If $\theta \in D(\mathcal{A})$ then $\theta = \sum f_i D_i$ and we have

$$\theta(Q) = \sum f_i D_i(Q) = Q \sum (f_i / x_i) \in QS.$$

Thus $f_i \in x_i S$.

The **Euler derivation** is $\theta_E = \sum x_i D_i$. If $f \in S_p$ is homogeneous of degree p then $\theta_E(f) = pf$. Thus θ_E is independent of the basis for V^*. Since $\theta_E(Q) = |\mathcal{A}| Q \in QS$, $\theta_E \in D_S(\mathcal{A})$ for all $\mathcal{A}$. In case S is fixed, it is convenient to write $D(\mathcal{A}) = D_S(\mathcal{A})$.

Note that $\theta \in \mathrm{Der}(S)$ may be written as $\theta = \sum \theta(x_i) D_i$. Given derivations $\theta_1, \ldots, \theta_\ell \in D(\mathcal{A})$ define a matrix $\mathsf{Q} = \mathsf{Q}(\theta_1, \ldots, \theta_\ell)$ by $\mathsf{Q}_{i,j} = \theta_j(x_i)$. Thus

$\theta_j = \sum \mathsf{Q}_{i,j} D_i$. Let $Q(\theta_1, \ldots, \theta_\ell) = \det \mathsf{Q}(\theta_1, \ldots, \theta_\ell)$. For $p, q \in S$ write $p \approx q$ if $p = cq$ for some $c \in \mathbf{K}^*$.

THEOREM 9.6. *If $\theta_1, \ldots, \theta_\ell \in D(\mathcal{A})$ then $Q(\theta_1, \ldots, \theta_\ell) \in QS$.*

THEOREM 9.7. *Suppose $\theta_1, \ldots, \theta_\ell \in D(\mathcal{A})$. The following conditions are equivalent:*
(i) *$Q(\theta_1, \ldots, \theta_\ell) \approx Q$,*
(ii) *$\mathcal{A}$ is free and $\{\theta_1, \ldots, \theta_\ell\}$ is a basis for $D(\mathcal{A})$.*

This allows us to prove that the braid arrangement is free. Recall that for the braid arrangement $Q = \prod_{1 \le i < j \le \ell} (x_i - x_j)$. Define for $1 \le p \le \ell$

$$\theta_p = \sum_{q=1}^{\ell} x_q^{p-1} D_q.$$

Note that

$$\theta_p(x_i - x_j) = x_i^{p-1} - x_j^{p-1} \in (x_i - x_j)S.$$

It follows from Corollary 9.3.i that $\theta_p \in D(\mathcal{A})$. Direct computation gives:

$$\mathsf{Q}(\theta_1, \ldots, \theta_\ell) = \begin{pmatrix} 1 & x_1 & \ldots & x_1^{\ell-1} \\ \vdots & \vdots & & \vdots \\ 1 & x_\ell & \ldots & x_\ell^{\ell-1} \end{pmatrix}.$$

Since this is the Vandermonde matrix we know that $Q(\theta_1, \ldots, \theta_\ell) \approx Q$. It follows from Theorem 9.7.ii that the braid arrangement is free and $\theta_1, \ldots, \theta_\ell$ is a basis for $D(\mathcal{A})$. Note that this basis consists of homogeneous derivations of degrees $-1, 0, 1, 2, \ldots, (\ell - 2)$.

THEOREM 9.8. *Let $\theta_1, \ldots, \theta_\ell \in D(\mathcal{A})$ be homogeneous derivations which satisfy:*
(i) *$\theta_1, \ldots, \theta_\ell$ are independent over S,*
(ii) *$\sum_{i=1}^{\ell} \deg \theta_i = |\mathcal{A}| - \ell$.*
Then $\mathcal{A}$ is free and $\theta_1, \ldots, \theta_\ell$ is a basis for $D(\mathcal{A})$.

THEOREM 9.9. *If $\mathcal{A}$ is free then*
(i) *$D(\mathcal{A})$ has a homogeneous basis $\theta_1, \ldots, \theta_\ell$,*
(ii) *the unordered ℓ-tuple $\{\deg \theta_1, \ldots, \deg \theta_\ell\}$ depends only on $\mathcal{A}$,*
(iii) *$\sum_{i=1}^{\ell} \deg \theta_i = |\mathcal{A}| - \ell$.*

DEFINITION 9.10. *If $\mathcal{A}$ is free call the unordered ℓ-tuple*

$$\deg \mathcal{A} = \{\deg \theta_1, \ldots, \deg \theta_\ell\}$$

the **degrees** *of $\mathcal{A}$ for any homogeneous basis $\theta_1, \ldots, \theta_\ell$ of $D(\mathcal{A})$.*

For example $\deg \Phi_\ell = \{-1, \ldots, -1\}$. For the Boolean arrangement $\deg \mathcal{A} = \{0, \ldots, 0\}$. For the braid arrangement

$$\deg \mathcal{A} = \{-1, 0, 1, 2, \ldots, (\ell - 2)\}.$$

PROPOSITION 9.11. *Let $\mathcal{A}$ be a 2-arrangement. Then*
(i) *$\mathcal{A}$ is free,*
(ii) *if $|\mathcal{A}| = n > 0$ then* $\deg \mathcal{A} = \{0, n-2\}$.

Proof. If Q defines $\mathcal{A}$ let $\theta_1 = \theta_E$ and let $\theta_2 = (D_2 Q)D_1 - (D_1 Q)D_2$. Then $\theta_1(Q) = (\deg Q)Q$ and $\theta_2(Q) = 0$ so $\theta_1, \theta_2 \in D(\mathcal{A})$ and

$$\mathsf{Q}(\theta_1, \theta_2) = \begin{pmatrix} x_1 & D_2 Q \\ x_2 & -D_1 Q \end{pmatrix}.$$

Assertion (i) follows from Theorem 9.7.i, and (ii) by counting degrees.

Next let us consider 3-arrangements and use x, y, z in place of x_1, x_2, x_3. If $|\mathcal{A}| \leq 3$ then suitable choice of coordinates gives defining polynomials $Q = 1$, $Q = x$, $Q = xy$, $Q = xyz$ and in each case it is easy to see that the corresponding arrangement is free. If $|\mathcal{A}| = 4$ the situation is more interesting. We may choose coordinates so that $Q = xyz(ax + by + cz)$ and at most one of a, b, c is zero. Thus there are two cases to consider.

(i) The arrangement $\mathcal{A}$ defined by $Q = xyz(ax + by)$ with $ab \neq 0$ is free. We may view this arrangement as a product $\mathcal{A} = \mathcal{A}_1 \times \mathcal{A}_2$ where $\mathcal{A}_1$ is the 1-arrangement defined by $Q_1 = z$, and $\mathcal{A}_2$ is the 2-arrangement defined by $Q_2 = xy(ax + by)$. Clearly $\mathcal{A}_1$ is free with basis $\theta_1 = zD_z$. It follows from Proposition 9.11 that $\mathcal{A}_2$ is free with basis $\theta_2 = xD_x + yD_y$, $\theta_3 = (ax^2 + 2bxy)D_x - (2axy + by^2)D_y$. Since $Q(\theta_1, \theta_2, \theta_3) \approx Q$, it follows from Theorem 9.7 that $\mathcal{A}$ is free. Note that $\deg \mathcal{A} = \{0, 0, 1\}$.

(ii) The arrangement $\mathcal{A}$ defined by $Q = xyz(ax + by + cz)$ with $abc \neq 0$ is not free. We argue by contradiction. Suppose $\mathcal{A}$ is free with $\deg \mathcal{A} = \{a_1, a_2, a_3\}$. We show first that $a_i \neq -1$. Suppose $\theta = AD_x + BD_y + CD_z \in D(\mathcal{A})$, where $A, B, C \in \mathbf{K}$. Since $\theta(x) = A$, $\theta(y) = B$ and $\theta(z) = C$ it follows from Corollary 9.3.i that $A = B = C = 0$. Thus $a_i \geq 0$. It follows from Theorem 9.9.iii that $a_1 + a_2 + a_3 = |\mathcal{A}| - \ell = 4 - 3 = 1$. Thus if $\mathcal{A}$ is free it must have $\deg \mathcal{A} = \{0, 0, 1\}$. Since $\theta_E \in D(\mathcal{A})$ and $\deg \theta_E = 0$ we obtain the required contradiction by showing that if $\theta \in D(\mathcal{A})$ with $\deg \theta = 0$ then θ is a constant multiple of θ_E. Write $\theta = pD_x + qD_y + rD_z$ where $p, q, r \in V^*$. Since $\theta(x) = p$, $\theta(y) = q$ and $\theta(z) = r$ it follows from Corollary 9.3.iii that $p = Ax$, $q = By$ and $r = Cz$ for $A, B, C \in \mathbf{K}$. Finally $\theta(ax + by + cz) = Aax + Bby + Ccz$ and Corollary 9.3.iii implies that $A = B = C$.

Consider the parameter space P spanned by a, b, c in this example. Then P is the subset of $\mathbf{R}^3$ where at most one coordinate is zero. Let S be the subset of P where one coordinate is zero. Then S is the parameter space of the free arrangements. The dense open set $P - S$ is the parameter space of the arrangements which are not free. This is a general fact. The condition for an arrangement to be free is algebraic and hence "non-generic". It is therefore quite surprising that so many of the arrangements which come from other applications are free.

We turn to consideration of subarrangements, restricted arrangements and triples. Suppose $\mathcal{B} \subset \mathcal{A}$. Then $Q(\mathcal{B}) = xyz$, $Q(\mathcal{A}) = xyz(x+y+z)$ shows that a free arrangement may be the subarrangement of one that is not free. To see that a subarrangement of a free arrangement is not necessarily free we assert that $\mathcal{A}$ defined by $Q(\mathcal{A}) = xyz(x+y)(x+y+z)$ is free and use $Q(\mathcal{B}) = xyz(x+y+z)$. To show that $\mathcal{A}$ is free would be laborious at this stage, since so far we can only determine when a given set $\theta_1, \ldots, \theta_\ell \in D(\mathcal{A})$ is a basis, and it is quite difficult to find *any* $\theta \in D(\mathcal{A})$ other than θ_E. We return to this example after Corollary 9.15.

THEOREM 9.12. *If $\mathcal{A}$ is free then $\mathcal{A}_X$ is free for all $X \in L(\mathcal{A})$.*

One of the most challenging conjectures in the subject is that if $\mathcal{A}$ is free then $\mathcal{A}^X$ is free for all $X \in L(\mathcal{A})$.

The Addition and Deletion Theorem

When $(\mathcal{A}, \mathcal{A}', \mathcal{A}'')$ is a triple the problem is delicate. We have just seen examples to show that it is possible for one of $\mathcal{A}$, $\mathcal{A}'$ to be free but not the other. First consider the case when both are free.

THEOREM 9.13. *Let $(\mathcal{A}, \mathcal{A}', \mathcal{A}'')$ be a triple such that $\mathcal{A}$ and $\mathcal{A}'$ are free. Then $\mathcal{A}''$ is free and there exist integers $a_1, \ldots, a_\ell$ such that:*

$$\begin{aligned} \deg \mathcal{A} &= \{a_1, \ldots, a_{\ell-1}, a_\ell\}, \\ \deg \mathcal{A}' &= \{a_1, \ldots, a_{\ell-1}, a_\ell - 1\}, \\ \deg \mathcal{A}'' &= \{a_1, \ldots, a_{\ell-1}\}. \end{aligned}$$

The most useful application of the triple $(\mathcal{A}, \mathcal{A}', \mathcal{A}'')$ is when information about $\mathcal{A}'$ and $\mathcal{A}''$ provides information about $\mathcal{A}$. Such an implication is given in the addition and deletion theorem:

THEOREM 9.14. *Let $(\mathcal{A}, \mathcal{A}', \mathcal{A}'')$ be a triple and assume that $\mathcal{A}''$ is free with* $\deg \mathcal{A}'' = \{a_1, \ldots, a_{\ell-1}\}$. *Then the following conditions are equivalent:*

(i) *$\mathcal{A}'$ is free with* $\deg \mathcal{A}' = \{a_1, \ldots, a_{\ell-1}, a_\ell - 1\}$,

(ii) *$\mathcal{A}$ is free with* $\deg \mathcal{A} = \{a_1, \ldots, a_{\ell-1}, a_\ell\}$.

Combined with Proposition 9.11 this gives:

COROLLARY 9.15. *Assume that $\ell = 3$ and $|\mathcal{A}''| = n > 0$. Then the following are equivalent:*

(i) *$\mathcal{A}'$ is free with* $\deg \mathcal{A}' = \{0, n-2, a-1\}$,

(ii) *$\mathcal{A}$ is free with* $\deg \mathcal{A} = \{0, n-2, a\}$.

This may be used to show that the 3-arrangement $\mathcal{A}$ defined by $Q(\mathcal{A}) = xyz(x+y)(x+y+z)$ is free. Let $H = \ker(x+y+z)$. We showed that $\mathcal{A}'$ defined by $Q(\mathcal{A}') = xyz(x+y)$ is free with $\deg \mathcal{A}' = \{0,0,1\}$. To find $|\mathcal{A}''| = n$ choose coordinates on H, say, $\tilde{x}, \tilde{y}$ so $z = -\tilde{x} - \tilde{y}$. Then $Q(\mathcal{A}'')$ is the product of the distinct factors obtained after substitution of $z = -\tilde{x} - \tilde{y}$ in $Q(\mathcal{A}')$. In our case $Q(\mathcal{A}'') = \tilde{x}\tilde{y}(\tilde{x} + \tilde{y})$ and $n = 3$. Corollary 9.15 implies that $\mathcal{A}$ is free with $\deg \mathcal{A} = \{0,1,1\}$. Note that this does not provide a basis for $D(\mathcal{A})$.

Inductively Free Arrangements

DEFINITION 9.16. *The class $\mathcal{IF}$ of* **inductively free** *arrangements is the smallest class of arrangements which satisfies:*

(i) $\Phi_\ell \in \mathcal{IF}$ *for* $\ell \geq 0$,

(ii) *If there exists $H \in \mathcal{A}$ such that $\mathcal{A}'' \in \mathcal{IF}$, $\mathcal{A}' \in \mathcal{IF}$ and* $\deg \mathcal{A}'' \subset \deg \mathcal{A}'$ *then $\mathcal{A} \in \mathcal{IF}$.*

It follows from an example in [139] that not all free arrangements are inductively free. Terao [147] showed that fiber type arrangements are inductively free. In order to show that a given arrangement $\mathcal{A}$ is inductively free, we must start with some inductively free arrangement and add hyperplanes one at a time satisfying (ii). This process may be described conveniently in an **induction table**. Each row is one step in the process. The first column gives the degrees of the arrangement which is the $\mathcal{A}'$ of that step. The second column gives α_H, where $H = \ker \alpha_H$ is the hyperplane added to $\mathcal{A}'$. The third column gives $\deg \mathcal{A}''$. The last row displays the degrees of $\mathcal{A}$. Since $Q(\mathcal{A}')$ is the product of the α_H in the rows above the row in consideration, it is easy to compute $Q(\mathcal{A}'')$. At each step the difficulty lies in showing that $\mathcal{A}''$ is free, and in computing $\deg \mathcal{A}''$. The induction table below illustrates the proof that $\mathcal{A}$ defined by $Q(\mathcal{A}) = xyz(x+y)(x+y+z)$ is inductively free. The most delicate problem is to determine in which order to add the hyperplanes. Even in this simple example the order of the last two hyperplanes could not be reversed.

$\deg \mathcal{A}'$	α_H	$\deg \mathcal{A}''$
$-1,-1,-1$	x	$-1,-1$
$-1,-1,0$	y	$-1,0$
$-1,0,0$	z	0,0
0,0,0	$x+y$	0,0
0,0,1	$x+y+z$	0,1
0,1,1		

The braid arrangement is also inductively free. This is argued by a double induction. For $k \leq \ell$ define an ℓ-arrangement $\mathcal{A}_\ell(k)$ by $Q_\ell(k) = \prod_{1 \leq i < j \leq k}(x_i - x_j)$. We will show that $\mathcal{A}_\ell(k)$ is inductively free with

$$\deg \mathcal{A}_\ell(k) = \{-1^{\ell-k+1}, 0, 1, \ldots, k-2\}.$$

Here we use a^m to denote the degree a repeated m times. By induction we may assume that $\mathcal{A}_p(q)$ is inductively free with the appropriate degrees for $p < \ell$, $q \leq p$, and that $\mathcal{A}_\ell(q)$ is inductively free with the appropriate degrees for $q < k$. We want to show that $\mathcal{A}_\ell(k)$ is inductively free. We may start with $\mathcal{A}_\ell(k-1)$ and add the hyperplanes $H_{i,k}$ for $1 \leq i < k$. The crucial fact is that $\mathcal{A}'' = \mathcal{A}_{\ell-1}(k-1)$ independently of i. Thus we get the following induction table:

$\deg \mathcal{A}'$	α_H	$\deg \mathcal{A}''$
$-1^{\ell-k+2}, 0, \ldots, k-3$	$x_1 - x_k$	$-1^{\ell-k+1}, 0, \ldots, k-3$
$-1^{\ell-k+1}, 0, 0, 1, \ldots, k-3$	$x_2 - x_k$	$-1^{\ell-k+1}, 0, \ldots, k-3$
$\vdots$	$x_i - x_k$	$-1^{\ell-k+1}, 0, \ldots, k-3$
$-1^{\ell-k+1}, 0, \ldots, k-3, k-3$	$x_{k-1} - x_k$	$-1^{\ell-k+1}, 0, \ldots, k-3$
$-1^{\ell-k+1}, 0, \ldots, k-2$		

In particular we recover the degrees of the braid arrangement

$$\deg \mathcal{A} = \{-1, 0, 1, \ldots, (\ell-2)\}.$$

These numbers differ only by one from the numbers that appear in the Poincaré polynomial $\pi(\mathcal{A}, t)$ of the braid arrangement in Proposition 2.26. This is a general fact.

Factorization Theorem

THEOREM 9.17. *Suppose* $\mathcal{A} \in \mathcal{IF}$ *with* $\deg \mathcal{A} = \{a_1, \ldots, a_\ell\}$. *Write* $b_i = a_i + 1$. *Then*

$$\pi(\mathcal{A}, t) = \prod_{i=1}^{\ell}(1 + b_i t).$$

Proof. We argue by induction on $|\mathcal{A}|$. If $|\mathcal{A}| = 0$ then $\mathcal{A} = \Phi_\ell$. Since $\pi(\Phi_\ell, t) = 1$ and $\deg \Phi_\ell = \{-1^\ell\}$, the assertion holds. For the induction step let $\mathcal{A}'$ and $\mathcal{A}''$ be inductively free with $\deg \mathcal{A}' = \{a_1, \ldots, a_{\ell-1}, a_\ell - 1\}$ and $\deg \mathcal{A}'' = \{a_1, \ldots, a_{\ell-1}\}$. By the induction hypothesis we have

$$\pi(\mathcal{A}', t) = \prod_{i=1}^{\ell-1}(1 + b_i t)(1 + a_\ell t),$$

$$\pi(\mathcal{A}'', t) = \prod_{i=1}^{\ell-1}(1 + b_i t).$$

In Corollary 2.29 we proved the formula:

$$\pi(\mathcal{A}, t) = \pi(\mathcal{A}', t) + t\pi(\mathcal{A}'', t).$$

Thus

$$\begin{aligned}\pi(\mathcal{A}, t) &= \prod_{i=1}^{\ell-1}(1 + b_i t)(1 + a_\ell t) + t\prod_{i=1}^{\ell-1}(1 + b_i t) \\ &= \prod_{i=1}^{\ell-1}(1 + b_i t)[1 + (a_\ell + 1)t] \\ &= \prod_{i=1}^{\ell}(1 + b_i t).\end{aligned}$$

By Theorem 9.14 this proves the assertion.

Definition 9.18. *Define the* **exponents** *of a free arrangement* $\mathcal{A}$ *with* $\deg \mathcal{A} = \{a_1, \ldots, a_\ell\}$ *by*

$$\exp \mathcal{A} = \{b_1, \ldots, b_\ell\}$$

where $b_i = a_i + 1$.

The terminology is derived from the result, see Theorem 10.7, that for a Coxeter group G and its reflection arrangement $\mathcal{A}(G)$ the exponents of G are the exponents of $\mathcal{A}(G)$. Terao [141] proved a stronger version of Theorem 9.17:

Theorem 9.19. *Suppose* $\mathcal{A}$ *is a free arrangement with* $\exp \mathcal{A} = \{b_1, \ldots, b_\ell\}$. *Then*

$$\pi(\mathcal{A}, t) = \prod_{i=1}^{\ell}(1 + b_i t).$$

This allows us to determine which discriminantal arrangements are free.

Proposition 9.20. *The discriminantal arrangements* $\mathcal{B}(k + 3, k)$ *are free if and only if* $k = 1$.

Proof. For $k = 1$ we have a braid arrangement, which is free. Write $\mathcal{B}_k = \mathcal{B}(k + 3, k)$. It suffices to show that if there exist integers a, b such that

$$\pi(\mathcal{B}_k, t) = (1 + t)(1 + at)(1 + bt)$$

then $k = 1$. If we factor the Poincaré polynomial given in Proposition 8.14 we get $\pi(\mathcal{B}_k) = (1 + t)p_k(t)$ where

$$p_k(t) = 1 + (1/2)(k^2 + 5k + 4)t + (1/8)(k^4 + 6k^3 + 15k^2 + 18k + 8)t^2.$$

The discriminant of $p_k(t)$ is

$$D(k) = -(1/4)k(k+1)(k^2+k-4).$$

Thus $D(1) = 1$ and $D(k) < 0$ for $k \geq 2$.

10 Reflection Arrangements

In the Introduction we defined reflections, reflecting hyperplanes, reflection groups and reflection arrangements. In this section we discuss some of their special properties.

Algebraic and Combinatorial Properties

We introduced the B_3-arrangement in Example 1.8. The corresponding reflection group is called the Coxeter group of type B_3. It is generated by reflections about the symmetry planes of the cube. The braid arrangement of Example 1.10 is also a reflection arrangement corresponding to the symmetric group. It is generated by the transpositions (i, j) with fixed set $H_{i,j} = \ker(x_i - x_j)$. These are examples of real reflection groups. They are also called Coxeter groups because finite irreducible real reflection groups were classified by Coxeter [26]. Shephard and Todd [130] classified finite irreducible complex reflection groups. Every real reflection group may be viewed as a complex reflection group. We give two examples of complex reflection groups which do not arise this way.

EXAMPLE 10.1. *The monomial groups $G(r, 1, \ell)$.*

Let $r \geq 2$ be an integer and let $C(r)$ be the cyclic group of order r generated by $\xi = e^{2\pi i/r}$. The group $G(r, 1, \ell)$ is the wreath product of $C(r)$ and $\mathrm{Sym}(\ell)$ consists of all $\ell \times \ell$ monomial matrices with entries in $C(r)$. Its reflection arrangement is defined by

$$Q = x_1 \cdots x_\ell \prod_{1 \leq i < j \leq \ell} (x_i^r - x_j^r).$$

Its lattice is the Dowling lattice $Q_\ell(\mathbf{Z}_r)$, see [36]. In particular $G(r, 1, 2)$ is generated by the matrices:

$$s_1 = \begin{pmatrix} \xi & 0 \\ 0 & 1 \end{pmatrix}, \quad s_2 = \begin{pmatrix} 0 & 1 \\ 1 & 0 \end{pmatrix}.$$

EXAMPLE 10.2. *The Hessian configuration.*

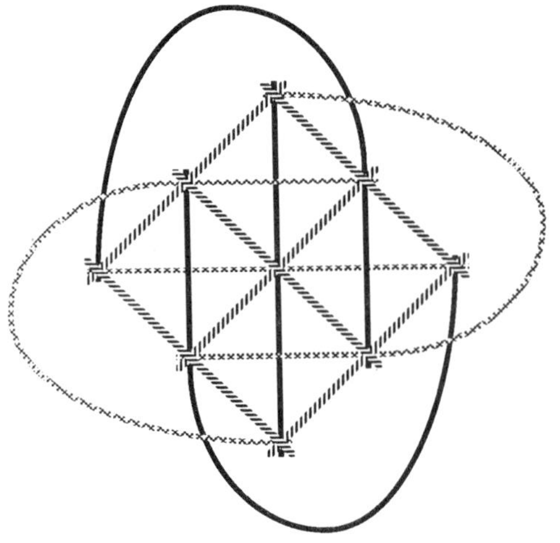

Figure 26: The Hessian configuration

Every nonsingular cubic in $\mathbf{C}P^2$ is projectively equivalent to one defined by $f(x, y, z) = x^3 + y^3 + z^3 - 3axyz$ with $a^3 \neq 1$ and $a \neq \infty$. The inflection points of a cubic are the solutions of $f = 0 = H(f)$, where $H(f)$ is the Hessian determinant of second partials. Direct calculation shows that for this family of curves the nine inflection points are independent of a. Set $\omega = e^{2\pi i/3}$. The projective coordinates of the nine inflection points are:

$$\begin{array}{lll} (0,1,-1) & (0,1,-\omega) & (0,1,-\omega^2) \\ (1,0,-1) & (1,0,-\omega) & (1,0,-\omega^2) \\ (1,-1,0) & (1,-\omega,0) & (1,-\omega^2,0) \end{array}$$

These 9 points lie on 12 projective lines, which are the four degenerate cubics corresponding to the parameter values $a = \infty$ and $a^3 = 1$:

$$x = 0, \quad y = 0, \quad z = 0, \quad x + \omega^i y + \omega^j z = 0$$

where $i, j = 0, 1, 2$. These 12 projective lines meet in 12 additional points. This configuration of 12 lines and 21 points is called the Hessian configuration. Each of the first nine points is contained in four lines, so we refer to them as quadruple points. Each of the second twelve points is contained in two lines, so we refer to them as double points. Each line contains three quadruple points and two double points. Figure 26 illustrates the configuration. The Hessian configuration has a distinguished history. For a complete account see Brieskorn and Knörrer's book [21]. It appeared recently in the work of Hirzebruch [67]. Its group of symmetries was determined by Jordan in 1878 as a subgroup of $PGL(2, \mathbf{C})$ of order 216. It is generated by projective transformations of order 3 which leave one of the 12 projective lines pointwise fixed. In addition, the group contains 9 projective transformations of order 2 which fix projective lines. This group gives rise to two complex reflection groups. One is called G_{25} in the classification of Shephard and Todd [130]. It has order 648. It is generated by reflections of order 3. Its reflection arrangement is defined by:

$$Q_1 = xyz \prod_{i,j=0,1,2} (x + \omega^i y + \omega^j z).$$

The other is called G_{26}. It has order 1296 and contains additional reflections of order 2. Its reflection arrangement is defined by:

$$Q_2 = Q_1(x^3 - y^3)(x^3 - z^3)(y^3 - z^3).$$

Next return to the general case of $G \subset GL(V)$. Let $\langle x, v\rangle = x(v)$ denote the usual pairing $V^* \times V \to \mathbf{K}$. Recall the S-module of derivations $\mathrm{Der}_S = \mathrm{Der}_{\mathbf{K}}(S)$ from Definition 1.17 and the S-module of differential 1-forms $\Omega_S = \Omega^1(V)$ from Definition 7.1. Let $\langle \omega, \theta\rangle = \omega(\theta)$ denote the natural pairing $\Omega_S \times \mathrm{Der}_S \to S$. If $v \in V$ let $D_v \in \mathrm{Der}_S$ be the derivation defined by $D_v(x) = \langle x, v\rangle$ for $x \in V^*$. If $f \in S$ then $df \in \Omega_S$ is defined by $\langle df, \theta\rangle = \theta(f)$. In terms of the bases D_i for Der_S and dx_j for Ω_S we have $df = \sum(D_i f)dx_i$ and $\theta(f) = \sum \theta(x_i)(D_i f)$.

The spaces S, Der_S, and Ω_S have G-module structures.

Definition 10.3. *Let $g \in G$, $v \in V$, $a \in S$, $\theta \in \mathrm{Der}_S$, and $\omega \in \Omega_S$. Define the G-module structure*

(i) *in S by $(ga)(v) = a(g^{-1}v)$,*
(ii) *in Der_S by $(g\theta)(a) = g(\theta(g^{-1}a))$,*
(iii) *in Ω_S by $(g\omega)(\theta) = g(\omega(g^{-1}\theta))$.*

Proposition 10.4. *The following transformation formulas hold:*

(i) $g(D_v a) = D_{gv}(ga)$, $d(ga) = g(da)$,
(ii) $g(D_v) = D_{gv}$, $g(a\theta) = (ga)(g\theta)$,
(iii) $g(a\omega) = (ga)(g\omega)$, $\langle g\omega, g\theta\rangle = g\langle \omega, \theta\rangle$.

Let $R = S^G$ be the ring of G-invariant polynomials. Let Der_S^G be the R-module of G-invariant derivations, and let Ω_S^G be the R-module of G-invariant differential forms. If $G \subset GL(V)$ is a reflection group then Chevalley's theorem [25] describes R, and [106, Lemma 2.21] describes Der_S^G and Ω_S^G.

Theorem 10.5. *Let $G \subset GL(V)$ be a finite reflection group.*

(i) *There exist homogeneous polynomials $f_1, \ldots, f_\ell$ such that*

$$R = \mathbf{K}[f_1, \ldots, f_\ell].$$

(ii) *The R-module* Ω_S^G *is free of rank* ℓ *with basis* $df_1, \ldots, df_\ell$.

(iii) *The R-module* Der_S^G *is free of rank* ℓ *with a basis of homogeneous elements.*

A set $\mathcal{F} = \{f_1, \ldots, f_\ell\}$ which satisfies (i) is called a set of **basic invariants** for G. The polynomials f_i are not unique, but their degrees $d_i = \deg f_i$ are determined uniquely by G. They are called the basic degrees. The integers $m_i = d_i - 1$ are the **exponents** of G. It is customary to label them in increasing order:

$$m_1 \leq \cdots \leq m_\ell.$$

A homogeneous basis for Der_S^G is called a set of **basic derivations** $\Theta = \{\theta_1, \ldots, \theta_\ell\}$. Let $a_i = \deg \theta_i$. The integers $n_i = a_i + 1$ are called the **coexponents** of G. It is customary to label them in increasing order:

$$n_1 \leq \cdots \leq n_\ell.$$

If G is a Coxeter group then $m_i = n_i$. Terao [140] proved the following:

THEOREM 10.6. *If* $\mathcal{A}$ *is a reflection arrangement then* $D(\mathcal{A}) = S \otimes_R \mathrm{Der}_S^G$.

Thus we get from Theorems 10.5, 10.6 and 9.19:

THEOREM 10.7. *If* $G \subset GL(V)$ *is a finite reflection group then its reflection arrangement* $\mathcal{A} = \mathcal{A}(G)$ *is free with*

$$\exp \mathcal{A} = \{n_1, \ldots, n_\ell\}.$$

The group G acts on the lattice $L(\mathcal{A})$. The orbits of this action were computed for irreducible reflection groups in [103] and [104]. In the course of this work we found the following :

PROPOSITION 10.8. *Let* $G \subset GL(V)$ *be a finite reflection group with reflection arrangement* $\mathcal{A} = \mathcal{A}(G)$. *For each* $X \in L(\mathcal{A})$ *with* $r(X) = p$ *there exist integers* $b_1^X, \ldots, b_p^X$ *such that*

$$\pi(\mathcal{A}^X, t) = \prod_{i=1}^{p} (1 + b_i^X t).$$

We demonstrate such a calculation for the group G_{25} defined in Example 10.2. For $v \in V$ let $G_v = \{g \in G | gv = v\}$ be the fixer of v. For $X \in L$ let $G_X = \bigcap_{v \in X} G_v$ be the fixer of X. The group G_{25} has five orbits on $L(\mathcal{A})$:

(i) V has fixer the identity group, called A_0,

(ii) the 12 hyperplanes form a single orbit with fixer $C(3)$,

(iii) the 12 lines which are in only two planes form an orbit with fixer $C(3) \times C(3)$,

(iv) the 9 lines which are in four planes form an orbit with fixer the group called G_4 in the classification,

(v) the origin is an orbit with fixer G_{25}.

The table below summarizes what we know about the lattice. The orbits are labeled by their fixers. The entry in row i column j is the number of elements in orbit j contained in an element in orbit i.

	$\mathcal{A}_0$	$C(3)$	$C(3)^2$	G_4	G_{25}	b_1^X	b_2^X	b_3^X
$\mathcal{A}_0$	1	12	12	9	1	1	4	7
$C(3)$		1	2	3	1	1	4	
$C(3)^2$			1	0	1	1		
G_4				1	1	1		
G_{25}					1			

It is clear how to compute b_1^X if X is a line in the orbit $C(3)^2$ or G_4. If X is a plane in the orbit $C(3)$ then $\mathcal{A}^X$ is a 2-arrangement with $|\mathcal{A}^X| = 5$ and hence $\pi(\mathcal{A}^X, t) = 1 + 5t + 4t^2 = (1+t)(1+4t)$. Finally, to compute $\pi(\mathcal{A}, t)$ note that if X is in the $C(3)$ orbit, then $\mu(X) = -1$, if X is in the $C(3)^2$ orbit then it is in two planes so $\mu(X) = 1$ and if X is in the G_4 orbit then it is in four planes so $\mu(X) = 3$. This allows calculation of $\mu(\{0\}) = -28$. Thus

$$\pi(\mathcal{A}, t) = 1 + 12t + 12t^2 + 27t^2 + 28t^3 = (1+t)(1+4t)(1+7t).$$

Proposition 10.8 motivated the conjecture that the restriction of a free arrangement is free. We would need to know not only that $\mathcal{A}^X$ is free but also that $\exp\mathcal{A}^X = \{b_1^X, \ldots, b_p^X\}$. In all the examples we have computed this is true but we have been able to prove it in only one case [110]:

THEOREM 10.9. *Let G be a finite Coxeter group with exponents $m_1 \leq \cdots \leq m_\ell$. Let $\mathcal{A} = \mathcal{A}(G)$ be its reflection arrangement. For $H \in \mathcal{A}$ the restriction $\mathcal{A}^H$ is free with*

$$\exp\mathcal{A}^H = \{m_1, \ldots, m_{\ell-1}\}.$$

Unfortunately, induction may not be used in combination with Theorem 10.9 because in general $\mathcal{A}^H$ is not the arrangement of any Coxeter group. An even stronger conjecture says that every restriction of a reflection arrangement is inductively free. We have succeeded in showing this in many cases. The induction table below proves that the arrangement of the group G_{25} defined in Example 10.2 is inductively free. The reader willing to experiment with this example will soon discover how sensitive it is to the order of the hyperplanes.

exp$\mathcal{A}'$	α_H	exp$\mathcal{A}''$
0,0,0	$x+y+z$	0,0
0,0,1	$x+y+\omega z$	0,1
0,1,1	$x+y+\omega^2 z$	0,1
0,1,2	z	0,1
0,1,3	x	1,3
1,1,3	$x+\omega y+z$	1,3
1,2,3	$x+\omega y+\omega z$	1,3
1,3,3	$x+\omega y+\omega^2 z$	1,3
1,3,4	y	1,4
1,4,4	$x+\omega^2 y+z$	1,4
1,4,5	$x+\omega^2 y+\omega z$	1,4
1,4,6	$x+\omega^2 y+\omega^2 z$	1,4
1,4,7		

The $K(\pi, 1)$ Problem

We close this section with a discussion of the topology of the complements of reflection arrangements. Brieskorn [20] generalized the constructions given in Remark 5.9 for the braid space. Let W be a finite Coxeter group and let $(\mathcal{A}, V)$ be its complexified reflection arrangement. Let $M = M(\mathcal{A})$. Let $p : V \to V/W$ be the orbit map. Note that W acts freely in M. Let $B = M/W$. The branch locus of p is called the discriminant locus. There is a natural embedding of B into $\mathbf{C}^\ell$ as follows. Let $\mathcal{F} = \{f_1, \ldots, f_\ell\}$ be a set of basic invariants for W. The map $\tau : V/W \to \mathbf{C}^\ell$ defined by

$$\tau(Wv) = (f_1(v), \ldots, f_\ell(v))$$

is a bijection. Since every reflection has order two, the polynomial $Q^2 = \prod_{H \in \mathcal{A}} \alpha_H^2$ is an invariant. We may define a polynomial $\Delta(T_1, \ldots, T_\ell; \mathcal{F})$, called the **discriminant**, in the indeterminates $T_1, \ldots, T_\ell$ and depending on $\mathcal{F}$ by:

$$\Delta(f_1, \ldots, f_\ell; \mathcal{F}) = \prod \alpha_H^2.$$

The discriminant locus

$$D = \{(z_1, \ldots, z_\ell) \in \mathbf{C}^\ell | \Delta(z_1, \ldots, z_\ell; \mathcal{F}) = 0\}$$

is the image under τ of N/W. Let $\pi = \tau p$. Then we have the commutative diagram:

$$\begin{array}{ccc} V & \supset & M = V - N \\ \pi \downarrow & & \downarrow \pi \\ \mathbf{C}^\ell & \supset & \tau B = \mathbf{C}^\ell - D. \end{array}$$

Since W is a Coxeter group, it has a presentation with generators $s_1, \ldots, s_\ell$ and relations

(1) $$s_i^2 = 1, \qquad 1 \le i \le \ell,$$

(2) $$s_i s_j s_i \cdots = s_j s_i s_j \cdots, \qquad i \neq j$$

where there are a certain number $q_{i,j} = q_{j,i} \ge 2$ terms on both sides of equation (2). Let A be the Artin group with generators $a_1, \ldots, a_\ell$ and relations

(3) $$a_i a_j a_i \cdots = a_j a_i a_j \cdots, \qquad i \neq j$$

where there are $q_{i,j} = q_{j,i}$ terms on both sides of the equation. It is understood that the $q_{i,j}$ in (2) and (3) are the same. The natural surjection $A \to W$ gives rise to an exact sequence

(4) $$1 \to N_W \to A \to W \to 1.$$

Brieskorn [19] proved that $\pi_1(B) = A$. It follows from Deligne's Theorem 5.15 that M is a $K(\pi, 1)$-space. Thus (4) is the nontrivial part of the homotopy exact sequence of the fibration $p : M \to B$.

If G is a complex reflection group with reflection arrangement $(\mathcal{A}, V)$ the construction is similar. For each $H \in \mathcal{A}$ let e_H be the order of the cyclic subgroup fixing H. Then $\prod \alpha_H^{e_H}$ is a G-invariant polynomial. Given a set of basic invariants $\mathcal{F}$, we define the discriminant $\Delta(T_1, \ldots, T_\ell; \mathcal{F})$ by

$$\Delta(f_1, \ldots, f_\ell; \mathcal{F}) = \prod_{H \in \mathcal{A}} \alpha_H^{e_H}.$$

We define the orbit map $p : V \to V/G$, the bijection $\tau : V/G \to \mathbf{C}^\ell$ and the discriminant locus as before, and let $\pi = \tau p$ to get the commutative diagram:

$$\begin{array}{ccc} V & \supset & M = V - N \\ \pi \downarrow & & \downarrow \pi \\ \mathbf{C}^\ell & \supset & \tau B = \mathbf{C}^\ell - D. \end{array}$$

It is conjectured that M and B are $K(\pi, 1)$-spaces for all complex reflection groups. We observed in Proposition 5.7 that every central 2-arrangement is $K(\pi, 1)$. Nakamura [97] showed that if G is an imprimitive complex reflection group then its arrangement is $K(\pi, 1)$. This holds in particular for the groups defined in Example 10.1. The fundamental groups $\pi_1(B)$ are not known for all G. They were computed for $\ell = 2$ by Bannai [7].

Regular complex polytopes were introduced by Shephard [129]. Their symmetry groups are irreducible complex reflection groups and we call them **Shephard groups**. There are finite irreducible complex reflection groups which are not Shephard groups. There are also Coxeter groups which are not Shephard groups. Coxeter [28] showed that every Shephard group admits a presentation with generating reflections $s_1, \ldots, s_\ell$ and relations

$$(5) \qquad s_i^{p_i} = 1, \qquad 1 \leq i \leq \ell,$$

$$(6) \qquad s_i s_j s_i \cdots = s_j s_i s_j \cdots, \qquad i \neq j$$

where $p_i \geq 2$ are integers and there are a certain number $q_{i,j} = q_{j,i} \geq 2$ terms on both sides of equation (6). Not all complex reflection groups admit such presentations.

To each Shephard group G we associate a Coxeter group W by replacing p_i by 2 in (5). The group W is uniquely determined by G up to isomorphism. Thus G and W are both finite quotients of the same Artin group A. We call them an associated pair of groups and write (G, W). In general G and W are neither subgroups nor quotient groups of each other. When both groups are in consideration we use notation like M_G, M_W, d_i^G, d_i^W, etc. The following is a consequence of the main result of [107]:

THEOREM 10.10. *Let G be a Shephard group and let W be the associated Coxeter group. There exist basic sets $\mathcal{F}_G = \{f_1^G, \ldots, f_\ell^G\}$ and $\mathcal{F}_W = \{f_1^W, \ldots, f_\ell^W\}$ such that their discriminant polynomials are equal:*

$$\Delta_G(T_1, \ldots, T_\ell; \mathcal{F}_G) = \Delta_W(T_1, \ldots, T_\ell; \mathcal{F}_W).$$

Thus with this choice of coordinates G and W have the same discriminant loci, $D_G = D_W$ and hence $B_G = B_W$. Since it follows from Theorem 5.15 of Deligne that B_W is a $K(\pi, 1)$ we get:

COROLLARY 10.11. *If G is a Shephard group then $\mathcal{A}(G)$ is $K(\pi, 1)$.*

The group $G = G_{25}$ introduced in Example 10.2 is a Shephard group. The associated Coxeter group W is of type $A_3 = D_3$. We illustrate Theorem 10.10 with the pair (G_{25}, D_3). Maschke [91, p.326] constructed certain homogeneous polynomials $C_6, C_9, C_{12}, \mathcal{C}_{12}$ of degrees 6,9,12 and 12, where $\mathcal{C}_{12} = Q_1$ defines $\mathcal{A}(G)$. Shephard and Todd [130, p.286] remarked that we may choose $\mathcal{F}_G = \{C_6, C_9, C_{12}\}$ as basic invariants for G. It follows from Maschke's work [91, p.326] that

$$\mathcal{C}_{12}^3 \approx (432C_9^2 - C_6^3 + 3C_6C_{12})^2 - 4C_{12}^3.$$

Since every reflection in G has order 3, $\prod \alpha_H^{e_H} = \mathcal{C}_{12}^3$. Thus we have:

$$(7) \qquad \Delta_G(T_1, T_2, T_3; \mathcal{F}_G) = (432T_2^2 - T_1^3 + 3T_1T_3)^2 - 4T_3^3.$$

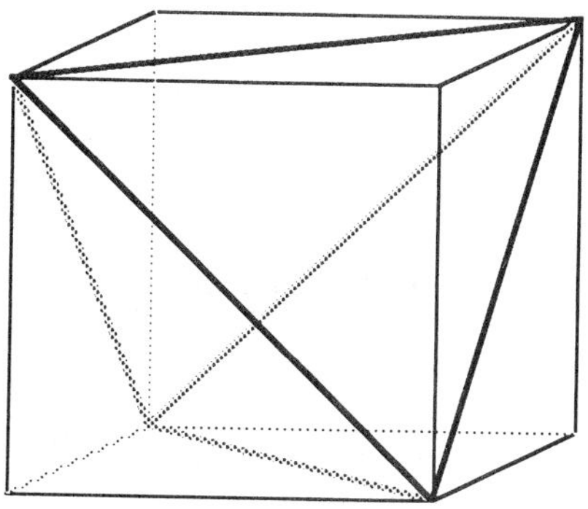

Figure 27: A tetrahedron in the cube

Next we compute the discriminant for $W = D_3$. Since D_3 is a subgroup of B_3 of index 2, we may think of W as the group of symmetries of one of the two tetrahedra inscribed in a cube, see Figure 27. Thus we may choose a basis x, y, z for V^* such that

$$\prod_{H \in \mathcal{A}(W)} \alpha_H = (x^2 - y^2)(x^2 - z^2)(y^2 - z^2).$$

In this coordinate system $p_1 = x^2 + y^2 + z^2$, $p_2 = xyz$, and $p_3 = x^2y^2 + x^2z^2 + y^2z^2$ is a set of basic invariants for W. Consider a cubic polynomial with roots x^2, y^2, z^2. The formula for the discriminant of this cubic, expressed in terms of the elementary symmetric functions of the roots, gives the identity

$$\prod_{H \in \mathcal{A}(W)} \alpha_H^2 = (2p_1^3 - 9p_1p_3 + 27p_2^2)^2 - 4(p_1^2 - 3p_3)^3.$$

Let $f_1 = p_1$, $f_2 = p_2/4$ and $f_3 = p_1^2 - 3p_3$. Then $\mathcal{F}_W = \{f_1, f_2, f_3\}$ is also a set of basic invariants for W and we have

$$\Delta_W(T_1, T_2, T_3; \mathcal{F}_W) = (432T_2^2 - T_1^3 + 3T_1T_3)^2 - 4T_3^3. \tag{8}$$

Comparing (7) and (8) illustrates Theorem 10.10 for the pair (G_{25}, D_3).

References

[1] M. Aigner, "Combinatorial Theory," Grundlehren der Math. Wiss. **234**, Springer-Verlag, Berlin/Heidelberg/New York, 1979.

[2] V. I. Arnold, Braids of algebraic functions and cohomologies of swallowtails, *Uspekhi Mat. Nauk* **23**(4) (1968), 247-248.

[3] ______, The cohomology ring of the colored braid group, *Mat. Zametki* **5** (1969), 227-231 : *Math. Notes* **5** (1969), 138-140.

[4] ______, Wave front evolution and equivariant Morse lemma, *Comm. Pure Appl. Math.* **29** (1976), 557-582.

[5] E. Artin, Theorie der Zöpfe, *Hamb. Abh.* **4** (1925), 47-72.

[6] K. Baclawski, Whitney numbers of geometric lattices, *Advances in Math.* **16** (1975), 125-138.

[7] E. Bannai, Fundamental groups of the spaces of regular orbits of the finite unitary reflection groups of dimension 2, *J. Math. Soc. Japan* **28** (1976), 447-454.

[8] M. Barnabei, A. Brini, and G.-C. Rota, The theory of Möbius functions, *Russian Math. Surveys* **41**(3) (1986), 135-188.

[9] G. Barthel, F. Hirzebruch, and T. Höfer, "Geradenkonfigurationen und Algebraische Flächen," Vieweg Publishing, Wiesbaden, 1987.

[10] M. Bayer and B. Sturmfels, Lawrence polytopes, preprint.

[11] G. Birkhoff, "Lattice Theory," 3rd ed., Amer. Math. Soc. Colloq. Publ., **25**, Providence, R.I., 1967.

[12] J. Birman, "Braids, links, and mapping classes," Annals of Math. Studies **87** Princeton Univ. Press, 1974.

[13] A. Björner, On the homology of geometric lattices, *Algebra Universalis* **14** (1982), 107-128.

[14] A. Björner, Homotopy type of posets and lattice complementation, *J. Comb. Theory (A)* **30** (1981), 90-100.

[15] A. Björner, P. Edelman and G. Ziegler, Hyperplane arrangements with a lattice of regions, preprint.

[16] A. Björner and J. W. Walker, A homotopy complementation formula for partially ordered sets, *Europ. J. Combinatorics* **4** (1983), 11-19.

[17] A. Björner and G. M. Ziegler, Broken circuit complexes: factorizations and generalizations, preprint.

[18] N. Bourbaki, "Groupes et Algèbres de Lie," Chapitres 4,5 et 6, Hermann, Paris, 1968.

[19] E. Brieskorn, Die Fundamentalgruppe des Raumes der regulären Orbits einer endlichen komplexen Spiegelungsgruppe, *Invent. Math.* **12** (1971), 57-61.

[20] E. Brieskorn, Sur les groupes de tresses, in "Séminaire Bourbaki 1971/72," Lecture Notes in Math. **317**, Springer-Verlag, Berlin/Heidelberg/New York, 1973, pp.21-44.

[21] E. Brieskorn and H. Knörrer, "Plane Algebraic Curves," Birkhauser, Boston, 1986.

[22] T. Brylawski, A decomposition for combinatorial geometries, *Trans. Amer. Math. Soc.* **171** (1972), 235-282.

[23] ———, The broken circuit algebra, *Trans. Amer. Math. Soc.* **234** (1977), 417-433.

[24] P. Cartier, Les arrangements d'hyperplans: un chapitre de géometrie combinatoire, in "Séminaire Bourbaki 1980/81," Lecture Notes in Math. **901**, Springer-Verlag, Berlin/Heidelberg/New York, 1981, pp.1-22.

[25] C. Chevalley, Invariants of finite groups generated by reflections, *Amer. J. Math.* **77** (1955), 778-782.

[26] H. S. M. Coxeter, Discrete groups generated by reflections, *Annals of Math.* **35** (1934), 588-621.

[27] ———, The product of the generators of a finite group generated by reflections, *Duke Math. J.* **18** (1951), 765-782.

[28] ———, "Regular Polytopes," 3rd ed., Dover, New York, 1973.

[29] ——, "Regular Complex Polytopes," Cambridge Univ. Press, 1974.

[30] H. Crapo, The Möbius function of a lattice, *J. Comb. Theory* **1** (1966), 120-131.

[31] H. Crapo and G.-C. Rota, "Combinatorial Geometries," MIT Press, Cambridge, MA, 1971.

[32] R. Deheuvels, Homologie des ensembles ordonnés et des espaces topologiques, *Bull. Soc. Math. France* **90** (1962), 261-321.

[33] P. Deligne, Les immeubles des groupes de tresses généralisés, *Invent. Math.* **17** (1972), 273-302.

[34] P. Deligne and G. D. Mostow, Monodromy of hypergeometric functions and non-lattice integral monodromy, *Publ. Math. IHES* **63** (1986), 5-89.

[35] A. Dold, "Lectures on Algebraic Topology," Springer-Verlag, Berlin/Heidelberg/New York, 1972.

[36] T. A. Dowling, A class of geometric lattices based on finite groups, *J. Comb. Theory (B)* **14** (1973), 61-86. Erratum, *ibid* **15** (1973), 211.

[37] P. Edelman, A partial order on the regions of $\mathbf{R}^n$ dissected by hyperplanes, *Trans. Amer. Math. Soc.* **283** (1984), 617-631.

[38] H. Esnault, Fibre de Milnor d'un cône sur une courbe plane singulière, *Invent Math.* **68** (1982), 477-496.

[39] E. Fadell and L. Neuwirth, Configuration spaces, *Math. Scand.* **10** (1962), 111-118.

[40] M. Falk, Geometry and topology of hyperplane arrangements, Ph.D. Thesis, University of Wisconsin-Madison, 1983.

[41] ——, Combinatorics and the singularity defined by a product of linear forms, preprint.

[42] ——, The minimal model of the complement of an arrangement of hyperplanes, *Trans. Amer. Math. Soc.*, **309** (1988), 543–556.

[43] ——, The cohomology and fundamental group of a hyperplane complement, in "Singularities," Contemporary Math., Amer. Math. Soc., to appear.

[44] ——, On the algebra associated with a geometric lattice, *Advances in Math.*, to appear.

[45] M. Falk and R. Randell, The lower central series of a fiber-type arrangement, *Invent. Math.* **82** (1985), 77-88.

[46] ———, On the homotopy theory of arrangements, in "Complex Analytic Singularities," Advanced Studies in Pure Math. **8**, North-Holland, 1987, pp.101-124.

[47] ———, The lower central series of generalized pure braid groups, in "Geometry and Topology," Lect. Notes in Pure and Appl. Math. **105**, Marcel Decker, New York, 1986, pp.103-108.

[48] ———, Braid groups and products of free groups, in "Braids," Contemporary Math. **78**, Amer. Math. Soc., 1988, pp.217-228.

[49] J. Folkman, The homology groups of a lattice, *J. Math. and Mech.* **15** (1966), 631-636.

[50] R. H. Fox and L. Neuwirth, The braid groups, *Math. Scand.* **10** (1962), 119-126.

[51] I. M. Gelfand, General theory of hypergeometric functions, *Soviet Math. (Doklady)* **33** (1986), 573-577.

[52] I. M. Gelfand and V. V. Serganova, Combinatorial geometries and torus strata on homogeneous compact manifolds, *Russian Math. Surveys* **42**(2) (1987), 133-168.

[53] I. M. Gelfand and A. V. Zelevinskii, Algebraic and combinatorial aspects of the general theory of hypergeometric functions, *Funct. Anal. and Appl.* **20** (1986), 183-197.

[54] M. Goresky and R. MacPherson, "Stratified Morse Theory," Springer-Verlag, Berlin/Heidelberg/New York, 1988.

[55] C. Greene, On the Möbius algebra of a partially ordered set, *Advances in Math.* **10** (1973), 177-187.

[56] ———, An inequality for the Möbius function of geometric lattices, *Studies in Appl. Math.* **54** (1975), 71-74.

[57] ———, The Möbius function of a partially ordered set, in "Ordered Sets," D. Reidel, 1982.

[58] C. Greene and T. Zaslavsky, On the interpretation of Whitney numbers through arrangements of hyperplanes, zonotopes, non-Radon partitions, and orientations of graphs, *Trans. Amer. Math. Soc.* **280** (1983), 97-126.

[59] P. GRIFFITHS AND J. MORGAN, "Rational homotopy theory and differential forms," Birkhauser, Boston, 1981.

[60] B. GRÜNBAUM, "Convex polytopes," Interscience, New York, 1967.

[61] ——, Arrangements of hyperplanes, in Proc. Second Louisiana Conf. on Combinatorics, Graph Theory, and Computing (Louisiana State Univ., Baton Rouge, La. 1971), pp. 41-106, Louisiana State Univ., Baton Rouge, La., 1971.

[62] ——, "Arrangements and spreads," CBMS Lecture Notes **10**, Amer. Math. Soc., 1972.

[63] B. GRÜNBAUM AND G. C. SHEPHARD, Simplicial arrangements in projective 3-space, *Mitt. Math. Sem. Univ. Giessen* **166** (1984), 49-101.

[64] A. HATTORI, Topology of C^n minus a finite number of affine hyperplanes in general position, *J. Fac. Sci. Univ. Tokyo* **22** (1975), 205-219.

[65] H. HENDRIKS, Hyperplane arrangements of large type, *Invent. Math.* **79** (1985), 375-381.

[66] M. HIRSCH, "Differential Topology," Springer-Verlag, Berlin/ Heidelberg/New York, 1976.

[67] F. HIRZEBRUCH, Arrangements of lines and algebraic surfaces, in "Arithmetic and Geometry," Vol. II, Progress in Math. **36**, Birkhauser, Boston, 1983, pp. 113-140.

[68] B. HUNT, Coverings and ball quotients with special emphasis on the 3-dimensional case, *Bonner Math. Schriften* **174**, 1986.

[69] M. JAMBU, Algèbre d'holonomie de Lie et certaines fibrations topologiques, *C. R. Acad. Sci. Paris* **306** (1988), 479-482.

[70] ——, Fiber-type arrangements and factorization properties, preprint.

[71] M. JAMBU AND L. LEBORGNE, Fonction de Möbius et arrangements d'hyperplans, *C. R. Acad. Sci. Paris* **303** (1986), 311-314.

[72] M. JAMBU AND H. TERAO, Arrangements libres d'hyperplans et treillis hyper-résolubles, *C. R. Acad. Sci. Paris* **296** (1983), 623-624.

[73] ——, Free arrangements of hyperplanes and supersolvable lattices, *Advances in Math.* **52** (1984), 248-258.

[74] ——, The broken-circuit algebra, preprint.

[75] T. KOHNO, On the minimal algebra and $K(\pi, 1)$-property of affine algebraic varieties, preprint.

[76] ———, Differential forms and the fundamental group of the complement of hypersurfaces, in "Singularities," Proc. Symp. Pure Math. **40** Part 1, Amer. Math. Soc., 1983, pp.655-662.

[77] ———, On the holonomy Lie algebra and the nilpotent completion of the fundamental group of the complement of hypersurfaces, *Nagoya Math. J.* **92** (1983), 21-37.

[78] ———, Série de Poincaré-Koszul associée aux groupes de tresses pures, *Invent. Math.* **82** (1985), 57-75.

[79] ———, Homology of a local system on the complement of hyperplanes, *Proc. of the Japan Acad. Ser. A* **62** (1986), 144-147.

[80] ———, Poincaré series of the Malcev completion of generalized pure braid groups, preprint.

[81] ———, Rational $K(\pi, 1)$ arrangements satisfy the LCS formula, preprint.

[82] ———, Holonomy Lie algebras, logarithmic connections, and the lower central series of fundamental groups, in "Singularities," Contemporary Math., Amer. Math. Soc., to appear.

[83] M. Las Vergnas, Convexity in oriented matroids, *J. Comb. Theory (B)* **29** (1980), 231-243.

[84] G. I. Lehrer, On the Poincaré series associated with Coxeter group actions on complements of hyperplanes, preprint.

[85] ———, On hyperoctahedral hyperplane complements, preprint.

[86] G. I. Lehrer and L. Solomon, On the action of the symmetric group on the cohomology of the complement of its reflecting hyperplanes, *J. Algebra* **104**(2) (1986), 410-424.

[87] A. Libgober, On the homotopy type of the complement to plane algebraic curves, *J. für Reine und Ang. Math.* **367** (1986), 103-114.

[88] S. MacLane, "Homology," Springer-Verlag, Berlin/Heidelberg/New York, 1963.

[89] Yu. I. Manin and V. V. Schechtman, Higher Bruhat orders related to the symmetric group, *Funct. Anal. and Appl.* **20** (1986), 148-150.

[90] ———, Arrangements of hyperplanes, higher braid groups and higher Bruhat orders, preprint.

[91] H. MASCHKE, Aufstellung des vollen Formensystems einer quaternären Gruppe von 51840 linearen substitutionen, *Math. Ann.* **33** (1888), 317-344.

[92] H. MATSUMURA, "Commutative Algebra," Benjamin/Cummings, Second Edition, 1980.

[93] J. MILNOR, "Singular points of complex hypersurfaces," Annals of Math. Studies **61**, Princeton University Press, 1968.

[94] N. E. MNËV, On manifolds of combinatorial types of configurations and convex polyhedra, Soviet Math. Dokl. **32** (1985), 335-337.

[95] J. MORGAN, The algebraic topology of smooth algebraic varieties, *Publ. Math. IHES* **48** (1978), 137-204.

[96] I. NARUKI, The fundamental group of the complement of Klein's arrangement of twenty-one lines, *Topology and Appl.*, to appear.

[97] T. NAKAMURA, A note on the $K(\pi, 1)$-property of the orbit space of the unitary reflection group $G(m, \ell, n)$, *Sci. Papers College of Arts and Sciences, Univ. Tokyo* **33** (1983), 1-6.

[98] NGUYÊÑ VIÊT DŨNG, The fundamental group of the space of regular orbits of the affine Weyl groups, *Topology* **22** (1983), 425-435.

[99] P. ORLIK, Basic derivations for unitary reflection groups, in "Singularities," Contemporary Math., Amer. Math. Soc., to appear.

[100] P. ORLIK AND L. SOLOMON, Combinatorics and topology of complements of hyperplanes, *Invent. Math.* **56** (1980), 167-189.

[101] ———, Unitary reflection groups and cohomology, *Invent. Math.* **59** (1980), 77-94.

[102] ———, Complexes for reflection groups, in "Algebraic Geometry," Lecture Notes in Math. **862**, Springer-Verlag, Berlin/Heidelberg/New York, 1981, pp.193-207.

[103] ———, Coxeter arrangements, in "Singularities," Proc. Symp. Pure Math. **40** Part 2, Amer. Math. Soc. 1983, pp.269-292.

[104] ———, Arrangements defined by unitary reflection groups, *Math. Annalen* **261** (1982), 339-357.

[105] ———, Arrangements in unitary and orthogonal geometry over finite fields, *J. Combinatorial Theory Ser. A* **38**(2) (1985), 217-229.

[106] ———, The Hessian map in the invariant theory of reflection groups, *Nagoya Math. J.* **109** (1988), 1-21.

[107] ——, Discriminants in the invariant theory of reflection groups, *Nagoya Math. J.* **109** (1988), 23-45.

[108] ——, Braids and discriminants, in "Braids," Contemporary Math. **78**, Amer. Math. Soc., 1988, pp.605-613.

[109] P. ORLIK, L. SOLOMON AND H. TERAO, Arrangements of hyperplanes and differential forms, in "Combinatorics and Algebra," Contemporary Math. **34**, Amer. Math. Soc., 1984, pp.29-65.

[110] ——, On Coxeter arrangements and the Coxeter number, in "Complex Analytic Singularities," Advanced Studies in Pure Math. **8** North-Holland, 1987, pp.461-477.

[111] ——, "Arrangements of hyperplanes," in preparation.

[112] F. PHAM, "Introduction a l'Étude Topologique des Singularités de Landau," Mémorial des Sci. Math. **164** Gauthier-Villars, Paris, 1967.

[113] D. QUILLEN, Homotopy properties of the poset of non-trivial p-subgroups of a group, *Advances in Math.* **28** (1978), 101-128.

[114] R. RANDELL, On the topology of non-isolated singularities, in "Geometric Topology," Academic Press, New York, 1979, pp.445-473.

[115] ——, On the fundamental group of the complement of a singular plane curve, *Quart. J. Math. Oxford* (2) **31** (1980), 71-79.

[116] ——, The fundamental group of the complement of a union of complex hyperplanes, *Invent. Math.* **69** (1982), 103-108. Correction *Invent. Math.* **80** (1985), 467-468.

[117] ——, Lattice-isotopic arrangements are topologically isomorphic, preprint.

[118] L. ROSE AND H. TERAO, private communication.

[119] G.-C. ROTA, On the foundations of combinatorial theory I. Theory of Möbius functions, *Z. Wahrscheinlichkeitsrechnung* **2** (1964), 340-368.

[120] ——, On the combinatorics of the Euler characteristic, in "Studies in Pure Math," presented to Richard Rado (L. Mirsky, ed.), Academic Press, London 1971, pp.221-233.

[121] K. SAITO, Regularity of Gauss-Manin connection of flat family of isolated singularities, in "Conference Notes," Centre de Mathématiques de l'Ecole Polytechnique, 1973.

[122] ——, On the uniformization of complements of discriminant loci, in "Conference Notes," AMS Summer Institute, Williamstown, 1975.

[123] ______, On a linear structure of a quotient variety by a finite reflexion group, *RIMS Kyoto preprint* **288**, 1979.

[124] ______, Theory of logarithmic differential forms and logarithmic vector fields, *J. Fac. Sci. Univ. Tokyo Sect. IA Math.* **27** (1981), 265-291.

[125] K. SAITO, T. YANO AND J. SEKIGUCHI, On a certain generator system of the ring of invariants of a finite reflection group, *Communications in Algebra* **8**(4) (1980), 373-408.

[126] M. SALVETTI, Topology of the complement of real hyperplanes in C^N, *Invent. Math.* **88** (1987), 603-618.

[127] ______, Arrangements of lines and monodromy of plane curves, preprint.

[128] ______, Generalized braid groups and self-energy Feynmann integrals, in "Braids," Contemporary Math. **78**, Amer. Math. Soc., 1988, pp.675-686.

[129] G. C. SHEPHARD, Regular complex polytopes, *Proc. London Math. Soc.* (3)**2** (1952), 82-97.

[130] G. C. SHEPHARD AND J. A. TODD, Finite unitary reflection groups, *Canad. J. Math.* **6** (1954), 274-304.

[131] L. SOLOMON AND H. TERAO, A formula for the characteristic polynomial of an arrangement, *Advances in Math.* **64** (1987), 305-325.

[132] A. SOMMESE, On the density of ratios of Chern numbers of algebraic surfaces, *Math. Ann.* **268** (1984), 207-221.

[133] E. SPANIER, "Algebraic Topology," McGraw-Hill, New York, 1966.

[134] R. P. STANLEY, Modular elements of geometric lattices, *Algebra Universalis* **1** (1971), 214-217.

[135] ______, Supersolvable lattices, *Algebra Universalis* **2** (1972), 214-217.

[136] ______, T-free arrangements of hyperplanes, in "Progress in Graph Theory," Academic Press, 1984, p.539.

[137] ______, "Enumerative Combinatorics," Vol. I, Wadsworth and Brooks/Cole, Monterey, CA, 1986.

[138] D. SULLIVAN, Infinitesimal computations in topology, *Publ. Math. IHES* **47** (1977), 269-331.

[139] H. TERAO, Arrangements of hyperplanes and their freeness. I,II, *J. Fac. Sci. Univ. Tokyo* **27** (1980), 293-320.

[140] ———, Free arrangements of hyperplanes and unitary reflection groups, *Proc. Japan Acad. Ser. A* **56** (1980), 389-392.

[141] ———, Generalized exponents of a free arrangement of hyperplanes and Shephard-Todd-Brieskorn formula, *Invent. Math.* **63** (1981), 159-179.

[142] ———, On Betti numbers of complements of hyperplanes, *Publ. RIMS Kyoto Univ.* **17** (1981), 567-663.

[143] ———, The exponents of a free hypersurface, in "Singularities", Proc. Symp. Pure Math. **40** Part 2, Amer. Math. Soc. 1983, pp.561-566.

[144] ———, Discriminant of a holomorphic map and logarithmic vector fields, *J. Fac. Sci. Univ. Tokyo Sect. IA Math.* **30** (1983), 379-391.

[145] ———, Free arrangements of hyperplanes over an arbitrary field, *Proc. Japan Acad. Ser.A* **59** (1983), 301-303.

[146] ———, The bifurcation set and logarithmic vector fields, *Math. Ann.* **263** (1983), 313-321.

[147] ———, Modular elements of lattices and topological fibration, *Advances in Math.* **62** (1986), 135-154.

[148] H. Terao and T. Yano, The duality of the exponents of free deformations associated with unitary reflection groups, in "Algebraic Groups and Related Topics," Advanced Studies in Pure Math. **6**, North-Holland, 1985, pp.339-348.

[149] A. N. Varchenko, Combinatorics and topology of the disposition of affine hyperplanes in real space, *Funct. Anal. and Appl.* **21** (1987), 9-19.

[150] ———, Morse theory on configurations of hyperplanes and periods of hypergeometric functions, preprint (in Russian).

[151] A. N. Varchenko and I. M. Gelfand, Heaviside functions of configurations of hyperplanes, *Funct. Anal. Appl.* **21** (1987), 255-270.

[152] B. L. van der Waerden, "Modern Algebra," Ungar, New York, 1950.

[153] L. Weisner, Abstract theory of inversion of finite series, *Trans. Amer. Math. Soc.* **38** (1935), 474-484.

[154] H. Whitney, A logical expansion in mathematics, *Bull. Amer. Math. Soc.* **38** (1932), 572-579.

[155] T. Yano and J. Sekiguchi, The microlocal structure of weighted homogeneous polynomials associated with Coxeter systems, I, *Tokyo J. Math.* **2**(2) (1979), 193-219; II *Tokyo J. Math.* **4**(1) (1981), 1-34.

[156] S. YUZVINSKY, Cohen-Macaulay seminormalizations of unions of linear subspaces, preprint.

[157] T. ZASLAVSKY, "Facing up to arrangements: Face-count formulas for partitions of space by hyperplanes," Memoirs Amer. Math. Soc., No. 154, 1975.

[158] _____, Counting the faces of cut-up spaces, *Bull. Amer. Math. Soc.* **81** (1975), 916-918.

[159] _____, Maximal dissections of a simplex, *J. Comb. Theory (A)* **20** (1976), 244-257.

[160] _____, A combinatorial analysis of topological dissections, *Advances in Math.* **25** (1977), 267-285.

[161] _____, Arrangements of hyperplanes; matroids and graphs, in "Proc. Tenth Southeastern Conf. on Combinatorics, Graph Theory and Computing (Boca Raton, 1979)," Vol. II, pp. 895-911.

[162] _____, The slimmest arrangements of hyperplanes: I. Geometric lattices and projective arrangements, *Geometriae Dedicata* **14** (1983), 243-259; II. Basepointed geometric lattices and Euclidean arrangements, *Mathematika* **28** (1981), 169-190.

[163] G. M. ZIEGLER, Algebraic combinatorics of hyperplane arrangements, Ph.D. Thesis, MIT, 1987.

[164] _____, The face lattice of hyperplane arrangements, *Discrete Math.*, to appear.

[165] _____, Multiarrangements of hyperplanes and their freeness, preprint.

[166] _____, Matroid representations and free arrangements, preprint.

[167] _____, Combinatorial construction of logarithmic differential forms, preprint.

$$\pi(\mathcal{A},t) \overline{} P(A,t)$$

$$\mathcal{A} \rightsquigarrow L(\mathcal{A}) \longrightarrow A(\mathcal{A})$$

$$\Big\downarrow \qquad\qquad ??$$

$$M(\mathcal{A}) \dashrightarrow H^*(M(\mathcal{A}))$$

$$Poin(M,t)$$

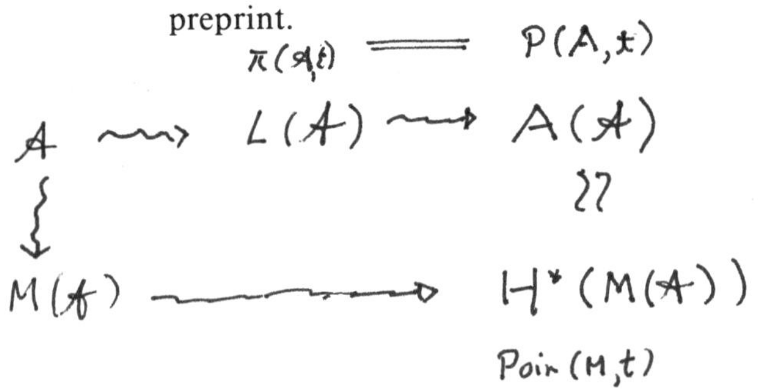

ABCDEFGHIJ — 89